AF588137

Global Environmental Studies

For further volumes:
http://www.springer.com/series/10124

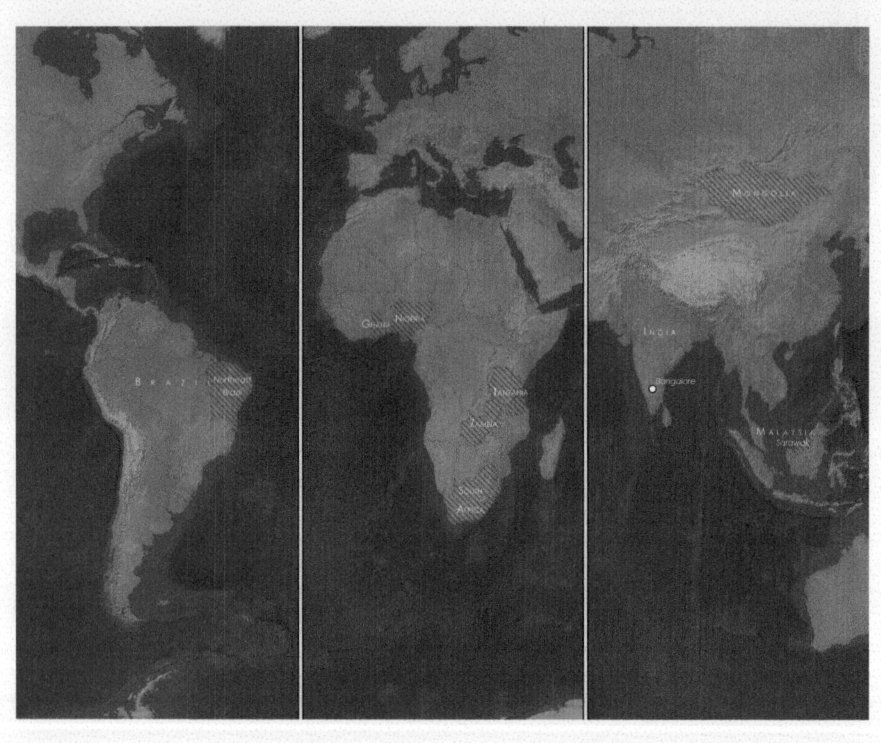
Ghana
Nigeria
Tanzania
Zambia
South
Africa
Mongolia
India
Bangalore
Malaysia
Sarawak

Shoko Sakai • Chieko Umetsu
Editors

Social–Ecological Systems in Transition

Editors
Shoko Sakai
Center for Ecological Research
Kyoto University
Otsu, Shiga, Japan

Chieko Umetsu
Graduate School of Fisheries Science
and Environmental Studies
Nagasaki University
Nagasaki, Japan

ISSN 2192-6336 ISSN 2192-6344 (electronic)
ISBN 978-4-431-54909-3 ISBN 978-4-431-54910-9 (eBook)
DOI 10.1007/978-4-431-54910-9
Springer Tokyo Heidelberg New York Dordrecht London

Library of Congress Control Number: 2014942559

Printed on acid-free paper

Springer is part of Springer Science+Business Media (www.springer.com)

Preface

Human activities have severely degraded most natural ecosystems, which are now in a critical condition. Various approaches have been developed to improve social–ecological systems for a better understanding of environmental problems and to explore better ways to increase the sustainability of both ecosystems and human societies. However, a clear perspective of how to address such problems is still lacking.

Part of the difficulty arises because of the diversity and complexity of ecosystems and human societies. Rich global biodiversity is a source of various ecological resources and services to human beings. Most terrestrial ecosystems affected by human activities comprise a mosaic of different land cover types or sub-ecosystems, and neighboring sub-ecosystems interact with one another through the migration of organisms and the flow of water and other materials. For example, recovery of biodiversity in secondary Bornean forests depends on the colonization of organisms from neighboring forests. In Mongolian grasslands, precipitation water is much better retained when there are adjacent forests. Because water availability is the primary determinant of plant growth, deforestation has a significant impact on grassland productivity and, thus, on the animals that depend solely on the vegetation.

In order to address environmental issues, actors in society who are involved in the land cover changes need to be identified. Some actors do not directly interact with the ecosystem but play a role indirectly via other actors. Ecological resources and services are utilized differently depending on the people, societies, and cultures. Even different actors in the same society use the same resources for different purposes, and this may sometimes cause conflicts. Primary forests used for timber production by international enterprises have been used for the collection of non-timber forest products by local people. Development of technologies and social and cultural changes have also altered the relationships between societies and ecosystems. Large-scale deforestation of primary forests in the tropics and transport of huge timbers that became possible about half a century ago make tropical timbers, which were useless for people living far from tropical areas, a highly valuable resource. On the other hand, the disappearance of natural resources in daily life motivates people living in cities to develop parks and plant trees in their neighborhood, even if it requires money, time, and land, although urban greenery might not have been given attention until recently. Because of diversity and complexity, predicting how effective a countermeasure would be to address an environmental problem or to foresee all possible outcomes before its implementation might not be easy.

The extremely rapid changes in the social and economical properties of social–ecological systems are another important factor. Recent drastic globalization of the economy and social relationships unite subsystems that might have been almost independent previously. Expansion and fusion of social–ecological systems are considered to basically destabilize the system, while the belief that globalization always worsens the situation might not be based on fact. Because of the rapid changes, even if a useful option to address a problem is found, it may be effective only for a limited time. One of the reasons that some researchers focusing on social–ecological systems emphasize adaptability and resilience rather than pursuing direct solutions is the great variability of the system.

Although there is no direct method to achieve high adaptability and resilience, one possible way is to compare and understand different problems associated with different social–ecological systems. The Research Institute for Humanity and Nature hosts international symposiums every year to discuss the relationships of humanity with nature from different points of view. Included in the contents are findings that came out of symposiums held in 2011 and 2012.

The book consists of ten chapters that have been divided into four parts. The first part, consisiting of a single chapter (Chapter 1), provides a broad prospective about what social–ecological systems are, with a review of the theoretical frameworks. Although different frameworks have been proposed and used in different backgrounds, an overarching theoretical framework is not yet available. We hope that some of the ideas discussed in this book may contribute to improving the situation.

The other chapters provide information about the specific aspects and problems in different social–ecological systems. The second part includes four chapters which discuss how human activities changed ecosystems from temperate grasslands to tropical areas. Although, in general, human activities are considered to decrease biodiversity, as shown in Chapter 2 regarding the changes in the species diversity of various biological groups, in the case of some ecosystems, maintenance of biodiversity is highly compatible with human activities, such as a grassland system under pastoralism, as shown in Chapters 3 and 4. Chapter 5 provides an example of how people reintroduce some natural resources and biodiversity into a mostly man-made environment.

The third part focuses on the adaptability of societies to unpredictable fluctuation in ecosystems. Although the present social–ecological systems may be more turbulent, information and knowledge regarding changes in the systems have been accumulated and institutions and techniques to cope with the unpredictability have been established (Chapter 6). Sustainable adaptation in the future should be addressed by analyzing current practices of climate change adaptation (Chapter 7).

The last part, with three chapters, is about factors for resilience of society against social and ecological shocks. Resilience is much more than recovery after a shock. It also involves the ability of individuals, communities, and entire regions to self-organize and increase their capacity for learning, experimentation, and adaptation. Chapters 8 and 9 examine factors related to resilience of relatively poor African communities in rural areas, which are very vulnerable to climatic fluctuations. Under similar physical environments, communities can differ in their adaptation strategies for increasing resilience. On the other hand, the last chapter, Chapter 10 introduces a story of recent changes in rural villages in Borneo, where large-scale development of plantations has altered land cover extensively. While some villages suffer from such development through a decrease in the opportunity for forest use, others take advantage of the development. Changes in resource use by local people seem to be flexible and adaptive, but the changes may potentially decrease resilience of the community by reducing the options they have.

The study of social–ecological systems is necessary to maintain our society. It is also a productive new field for discovering relationships between human society and nature. We hope this book will stimulate readers to do further research on this important and interesting research subject.

Otsu, Shiga, Japan — Shoko Sakai
Nagasaki, Japan — Chieko Umetsu

Contents

Global Environmental Studies

The Global Environmental Studies series introduces the research undertaken at, or in association with, the Research Institute for Humanity and Nature (RIHN). Located in Kyoto, Japan, RIHN is a national institute conducting fixed-term, multidisciplinary, international research projects on pressing areas of environmental concern.

RIHN seeks to transcend the common divisions between the humanities and the social and natural sciences, and to develop synthetic and transformative descriptions of humanity in the midst of a dynamic, changeable nature. The works published in the series will reflect the full breadth of RIHN scholarship in this transdisciplinary field of global environmental studies.

Editorial Structure

Part I
Frameworks and Concepts

Chapter 1
Theoretical Frameworks for the Analysis of Social–Ecological Systems

Graeme S. Cumming

Abstract Although the growing field of research on social–ecological systems (SESs) deals with some of the most important questions of our time, the study of SESs lacks an overarching theoretical framework. The development of such a framework is desirable because it would greatly improve our ability to generalize from individual case studies, to distinguish important from less important results, and ultimately to draw on the power of the scientific method to predict the consequence of management and policy interventions and to build greater resilience in SESs. Existing frameworks for the analysis of SESs can be grouped into five categories: (1) hypothesis-oriented frameworks; (2) assessment-oriented frameworks; (3) action-oriented frameworks; (4) problem-oriented frameworks; and (5) theory-oriented or overarching frameworks. Focusing on the fifth category, theory-oriented frameworks, seven assessment criteria are proposed that a satisfactory framework should meet: (1) it should provide a clear way of linking social and ecological systems and be strong in both disciplines; (2) it should be supported by rigorous empirical studies, key theories should meet Popper's falsifiability criterion, and frameworks should include translation modes that allow theory to be connected to empirical observations, and vice-versa; (3) frameworks should offer insights into causality, ideally being based on first principles, and should offer clear statements of cause and effect; (4) frameworks should deal with the dynamic aspects of SESs and the nature of change through time, as well as with the spatial nature of SES and spatial variation; (5) frameworks should build on previous frameworks and, ideally, should be able explain their weaknesses and/or incorporate their strengths; (6) frameworks for SESs should be able to cope with, and offer connections between, complementary perspectives and different epistemologies; and (7) frameworks should provide direction for the study of SESs by suggesting or

G.S. Cumming (✉)
Percy FitzPatrick Institute, University of Cape Town, Cape Town, South Africa
e-mail: gscumming@gmail.com

S. Sakai and C. Umetsu (eds.), *Social–Ecological Systems in Transition*, Global Environmental Studies, DOI 10.1007/978-4-431-54910-9_1, © Springer Japan 2014

guiding new empirical studies that will advance our theoretical understanding of SESs. Illustrative examples are offered from eight existing frameworks that meet some of these criteria, but highlight the fact that no existing framework meets all the criteria well. The development of a stronger theoretical framework remains an important goal for SES theory.

Keywords Framework • Philosophy • Resilience • Social–ecological system • Sustainability • Vulnerability

1.1 Introduction

The solutions to many of modern society's most pressing problems do not fall cleanly within the boundaries of a single discipline. Climate change, for example, has (as a minimum) geographic, ecological, political, sociological, and economic dimensions; understanding and resolving the problems that it presents requires data and collaboration across disciplines (IPCC 2007a, b). Similarly, the social dynamics of politics and power are among the primary drivers of habitat and species loss (e.g., Robbins 2004; Koontz and Bodine 2008; Johns 2009); the search for profit drives overfishing and deforestation; and ecosystem goods and services are increasingly being recognized as both exhaustible and profoundly important for human wellbeing (e.g., Folke et al. 1996; Costanza et al. 1999). If we use "Social–Ecological System" or "SES" to describe an integrated system of people and nature in which feedbacks occur between human and biophysical system elements (e.g., see Berkes et al. 2003; Adger et al. 2005; Ostrom 2009), and if we accept that most global problems are both social and ecological in form, then it is clear that the science of social-ecological systems will be fundamentally important for our collective future.

Curiously enough, in light of its importance, it is still debatable whether we have a science of SESs. The growing body of scientific theory relating to SESs (e.g., Folke 2006) draws heavily on systems ecology and complexity theory, but studies of SESs include some central societal concerns (for example, equity and human wellbeing) that have traditionally received little attention in complexity theory, and other contributing theories that deal with human society or human nature (e.g., game theory, behavioral psychology, or cultural theory) tend to be weak on ecological and system dynamic elements. Large-scale interdisciplinary and trans-disciplinary collaborations (e.g., Millennium Assessment 2003, 2005) are becoming increasingly common as scientists and governments gradually recognize that eco-systems influence human wellbeing, people are modifying or destroying ecosystems in many parts of the world, and the solutions to problems involving both people and nature require a holistic perspective on social–ecological systems (Carpenter 2002). Something that most large-scale collaborative ventures have made abundantly clear is that, although we are gradually defining a set of loosely related ideas and concepts relating to SESs, we still lack a cohesive body of SES theory.

The general lack of theoretical cohesion in studies of SESs makes it difficult, among other things, to identify generalities within and between case studies and to decide on the conceptual significance of isolated scientific findings. The power of science comes from its ability to find commonalities and connections between seemingly different phenomena and examples. If a greater level of generality in the analysis of SESs can be developed, we should be more able to understand how different case studies inform one another, how individual cases can advance theoretical understanding, and where assumptions and hypotheses are most in need of rigorous testing. This would help the field to advance from the level of proposing management heuristics to the level of making concrete, testable predictions.

This chapter explores the need for a theoretical framework for the study of SESs, the criteria that a framework should meet, and the elements of such a framework that are currently in place. An attainable theoretical framework for SESs may not be as all-encompassing as the unification of physics by the theories of relativity and thermodynamics, or the unification of biology by the theory of evolution, but it should play an important role in research on SESs (and the development of management and policy approaches for SESs) by organizing our thinking and both explaining and predicting important properties and behaviors of SESs.

I begin by confronting the difficult topic of reconciling epistemologies (ways of knowing). Adopting a scientific (empirical, rational, evidence-based) epistemology, I then propose a simple typology that describes the different types of analytical framework that currently exist for SESs. Narrowing down further from the typology, I focus on a single type of framework, the theory-oriented framework, and present eight *a priori* criteria that a theoretical framework for the analysis of SESs should strive to meet. Some published frameworks meet some of the different criteria, although no existing framework meets all of them, and I present a series of examples of frameworks that appear to contribute usefully to the development of a more general theory in one or more ways. I conclude by discussing areas in which further theoretical development and unification would be beneficial.

1.2 Reconciling Epistemologies

Before discussing theoretical frameworks, it is important to be clear on what constitutes what we will accept as "knowledge" and as "theory." The scientific study of SESs seeks to unite different disciplines by reference to a deeper, more abstract set of shared concepts and principles. For example, the concepts of diversity and selection can be applied to ecological systems (biodiversity and natural selection) and social systems (e.g., linguistic diversity and the loss of languages). Reference to this deeper set of general concepts does not, however, resolve the problem of differences in epistemologies. Students of the biophysical sciences are trained to expect empirical consistency and repeatability in the behavior of study systems, with the ultimate goal of prediction. Scientific understanding of the motions of the planets and the behavior of gases, and our ability to predict such phenomena as eclipses, offer a good example of the power of this approach to knowledge. Students of humans, in contrast, have long acknowledged that the complexities of human nature make formal scientific epistemologies virtually impossible to apply to many kinds of social science questions. Even neoclassical economics, which has access to huge amounts of data, must make unrealistic assumptions about rational choice and idealized markets in order to produce a consistent analytical framework for human economic behavior. Deeper reflection on human nature and the action of studying something suggests further complications: understanding has a subjective element and reality can be interpreted, or "constructed," in different ways. Knowledge claims that are based on empirical observation can in many cases be disproven by a combination of experimentation and logic, but claims of knowledge about human behavior are often impossible to disprove and must be assessed and tested according to a different set of disciplinary criteria. The nature of what is accepted as knowledge thus differs substantially between different disciplines.

An additional and fundamental problem in analyzing social–ecological systems is one the social sciences have grappled with for many years: any theoretical framework with relevance to explaining problems in a real-world SESs must incorporate not only observed patterns and processes of interest, but also a self-conscious analysis of the research process, because in a social system the process of research influences the conclusions. Agrawal and Gibson (1999) have argued that "…community must be examined in the context of development and conservation by focusing on the multiple interests and actors within communities, on how these actors influence decision-making, and on the internal and external institutions that shape the decision-making process." Such analysis can only be undertaken by engagement of the researcher with relevant communities. Similarly, Adger and Jordan (2009) make it clear that, in the context of governance, sustainability has "at least two important dimensions… the first is concerned with outcomes, the second with processes." As they explain, the processes by which decisions are made directly influence their outcomes. These considerations suggest that a theoretical framework for SESs must not only bridge epistemologies but also confront the relationships between methods and results.

It is unclear whether a successful reconciliation between different epistemologies will ever be possible in the study of SESs. Will social systems always be studied in different ways from ecological systems, for example, with social-ecological analyses providing a fertile meeting ground but not a single framework? Is the development of a single framework for the analysis of social–ecological systems desirable, or will progress be better served by the simultaneous development of different views and perspectives that will strengthen one another through their contrasting perspectives? I do not think it is currently possible to answer these questions in any definitive way, but I see several clear advantages in the development of a more cohesive theoretical framework for SESs. These would potentially include: (1) the development of better standards and more effective ways of assessing the quality of SES research, increasing rigor in analyses of SESs; (2) the creation of clearer linkages from the specific to the general, with case studies contributing more obviously to theoretical advancement; and (3) the development of better translation modes using theoretical constructs to generate evidence-based recommendations for social-ecological interventions which would enhance desirable aspects of social-ecological resilience.

Keeping these considerations in mind, in what follows I have focused on approaches that are based primarily in a scientific epistemology that assumes the primacy of empirical data and logic.

1.3 What Is a Scientific Framework?

According to Pickett et al. (2007), the goal of scientific theory is to facilitate understanding. Understanding in science can be defined as "an objectively determined, empirical match between some set of confirmable, observable phenomena… and a conceptual construct". Analysis, as a philosophical goal, can be considered as "the provision of necessary and sufficient conditions for a concept, or for the possession or application of a concept" (Glymour et al. 2009).

The main components of any scientific theory include domain, assumptions, concepts, definitions, facts, confirmed generalizations, laws, models, translation modes, hypotheses, and frameworks (Pickett et al. 2007). These different elements of theory are additive and interdependent. For example, defining the domain allows the scientist to determine the relevance of a related question or concept; and assumptions and definitions are subsequently used to explain ideas about cause and effect which are ultimately captured as models and laws.

For science to progress, the concepts comprising a body of theory must be explicitly connected to one another (Pickett et al. 2007). Connections are achieved through a set of overarching ideas giving them their significance. These ideas comprise the framework. A framework can be viewed as a family of models, and does not necessarily depend on deductive logic to connect different ideas (i.e., it does not have to present a single argument in which the conclusions follow from the premises). Most theoretical frameworks for SESs are hierarchical (rather than merely recognizing that SESs themselves are hierarchical), with different levels of the same theory explaining different aspects of the study system at different levels of generality. For example, a framework considering SESs as an interacting system of people and nature may incorporate sub-modules which focus primarily on social aspects of the system, such as decision-making or social networks.

Amidst the current haze of heuristics and case studies, we lack a unifying framework for the analysis of SESs—Kuhn's "disciplinary matrix" (Kuhn 1962)—that helps us to separate concepts, questions, and findings of genuine interest from background noise. Such a framework will firmly establish what we already know, highlight particular questions as being of particular interest, stimulate further investigations, and help the study of social–ecological systems to develop gradually into a more mature and more useful body of knowledge.

Although we do not currently have a unified theory of SESs, many relevant theories and frameworks have been proposed, and it is possible that a more unified theory (or subset of theories) may emerge from combining and building on the better elements of these different frameworks. The problem is further confused, however, by there being different kinds of frameworks in SES research and frameworks with different objectives which may be incommensurable. A typology of frameworks is offered here in response to this problem, as a way of clarifying the distinctions between their objectives and ensuring that we compare like with like.

1.4 A Typology of Frameworks for Analyzing SESs

Existing frameworks for the analysis of complex systems come in different forms, reflecting differences in the state of knowledge and different understandings of what constitutes a framework. Published frameworks (or at least texts that claim to be frameworks) range from systems of equations derived from first principles—as in the theory of relativity, for example—through to relatively simple box and arrow diagrams that explain quite loosely how different pieces of a system are expected to fit together. They may also be developed for different reasons. For example, Gibson et al. (2005) state that "frameworks are metatheoretical schema facilitating the organization of diagnosis, analysis, and prescription." While a good framework will undoubtedly facilitate the solution of problems, this definition comes equipped with a prescriptive agenda that may or may not be appropriate in a given situation.

Frameworks are developed and applied for different purposes. They are never "right" or "wrong," but they can be critically assessed using such criteria as their comprehensibility, their empirical rigor, and the degree to which they fulfill their purpose. Frameworks are developed for definite purposes. Frameworks with fundamentally different goals are incommensurable, and hence cannot be compared directly, although they may of course be linked to produce a multi-objective framework. It is thus important when we contrast frameworks, and when we evaluate the abilities of existing frameworks to contribute to the study of SESs, we compare like with like.

As I will explain in the next section, frameworks for social–ecological systems can be broadly grouped into five categories: (1) hypothesis-oriented frameworks; (2) assessment-oriented frameworks; (3) action-oriented frameworks; (4) problem-oriented frameworks; and (5) theory-oriented or overarching frameworks. I will discuss these categories briefly and then focus in detail on theory-oriented frameworks.

1.4.1 Hypothesis-Oriented Frameworks

These frameworks are quite specific, focusing on pairs of variables or clearly defined theoretical questions. An example of a framework in this category is the landscape disturbance framework of M. Turner et al. (2001), which is reproduced in

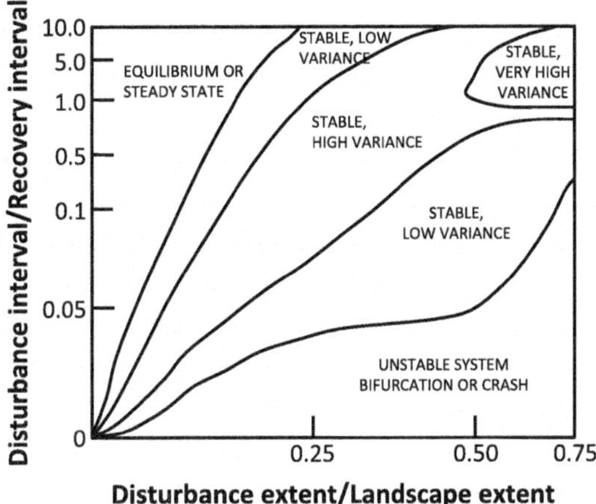

Fig. 1.1 An example of a hypothesis-oriented framework: the landscape disturbance framework, here redrawn from Monica Turner et al. (2001). The diagram suggests that ecological stability within a landscape depends on the relationship between the rate of disturbance and its extent; infrequent, localized disturbances promote stability, while frequent, regional disturbances promote instability

Fig. 1.1. It proposes a set of relationships between the size of an area that is disturbed and the time the ecological community in that area takes to recover, and translates easily into operational hypotheses and empirical tests. Some other examples of hypothesis-oriented frameworks include the intermediate disturbance hypothesis (Connell 1978), the application of neutral landscape models to explore relationships between deforestation and patch dynamics, the Kuznets curve in economics (which proposes a hump-shaped relationship between wealth and environmental protection; Kuznets 1955; Gergel et al. 2004), and Stommel diagrams (which display systemic properties or actors on spatial and temporal axes simultaneously; Stommel 1963).

1.4.2 Assessment-Oriented Frameworks

Assessment frameworks help people to think in a structured way about a system but are relatively mechanism-free. They are used to summarize key attributes of a social-ecological system for the purpose of describing it, typically for stock-taking or evaluative exercises. A good example of a widely used assessment framework is that of the Millennium Ecosystem Assessment (2003; Fig. 1.2). It consists of four boxes that capture the fundamental philosophy underpinning the MA: namely, human wellbeing and ecosystem services are intricately linked.

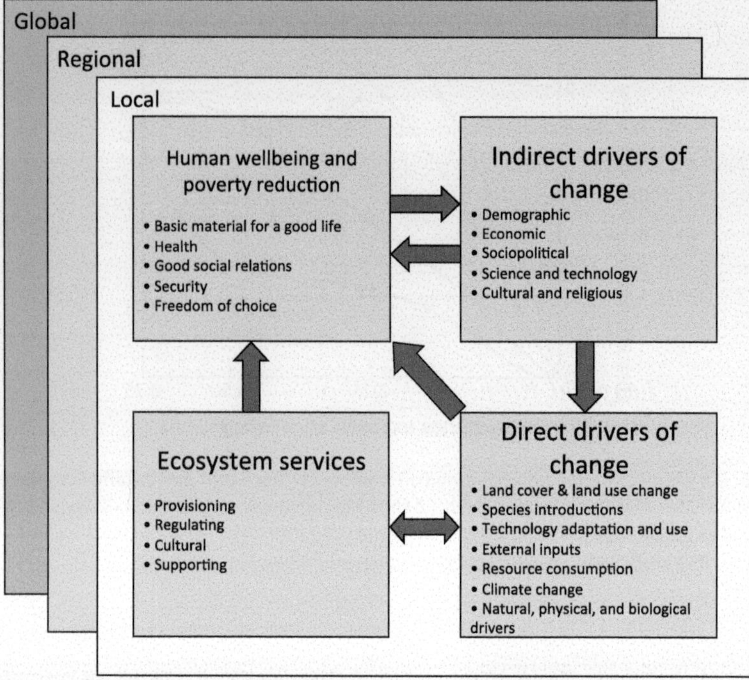

Fig. 1.2 An example of an assessment-oriented framework: a redrawn, slightly simplified version of the Millennium Ecosystem Assessment Framework

The Vulnerability Assessment Framework of B. Turner et al. (2003a, b; Fig. 1.3) is another important assessment-oriented framework. It summarizes social-ecological systems as consisting of: (1) a time- and location-specific, coupled human-environment system with distinctive features and levels of exposure, sensitivity, and coping capacity; (2) perturbations and disturbances impacting the local system; and (3) interactions with the broader (regional and global) social and environmental background. While the vulnerability framework offers a useful descriptive framework, like the MA framework, it lacks the active consideration of mechanisms needed for it to qualify as a fully blown theoretical framework (although aspects of vulnerability are increasingly being considered in theoretical terms; e.g., see Adger 2006).

1.4.3 Action-Oriented Frameworks

These recommend a particular course of action by an established set of actors in response to a particular kind of problem. They are usually focused on implementing solutions rather than establishing the causes of problems. Examples include The

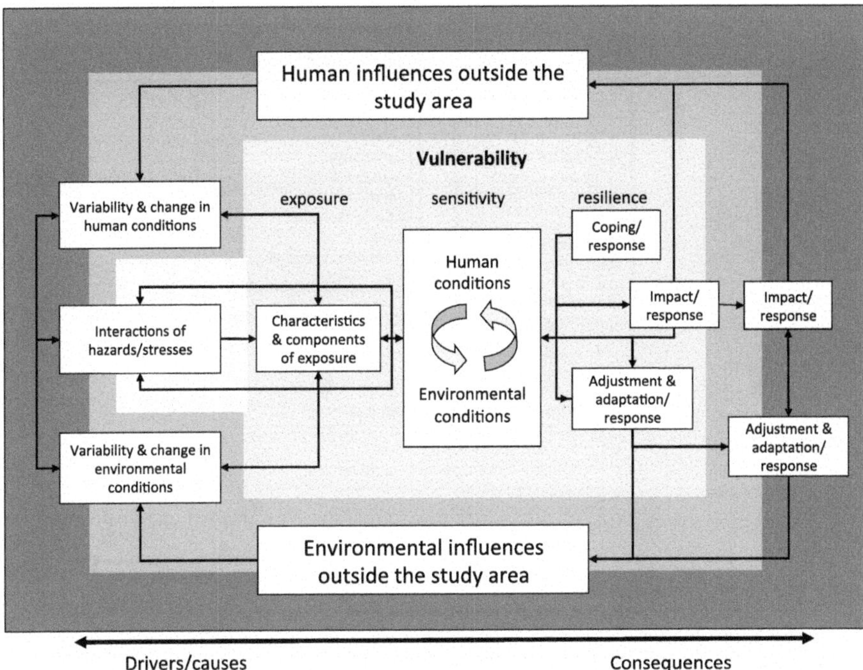

Fig. 1.3 A second example of an assessment-oriented framework: a redrawn version of the Vulnerability Framework of Billy-Lee Turner et al. (2003a, b)

Nature Conservancy's 5-S framework for conservation (TNC 2003) and the DPSIR (Driving forces-Pressure-State-Impacts-Responses) framework for conceptualizing the dynamics of invasive plant species (Fig. 1.4, Roura-Pascual et al. 2009). In some but not all cases, action-oriented frameworks are informed by a "command and control" mentality which assumes a problem is bounded and a feasible, direct solution exists (Holling and Meffe 1996); they may periodically need to be reassessed from a problem-oriented perspective.

1.4.4 Problem-Oriented Frameworks

Problem-oriented frameworks have mostly been developed to initiate and facilitate the *process* of solving a particular kind of problem. They focus more on problem identification and problem-solving processes than on prescribing the actual actions that are to be undertaken; unlike action-oriented frameworks, they do not start with a clear definition of the solution. It is important to note that problem-oriented frameworks are usually more concerned with the application of theory in particular instances, through collaborative processes, than with the advancement of theory

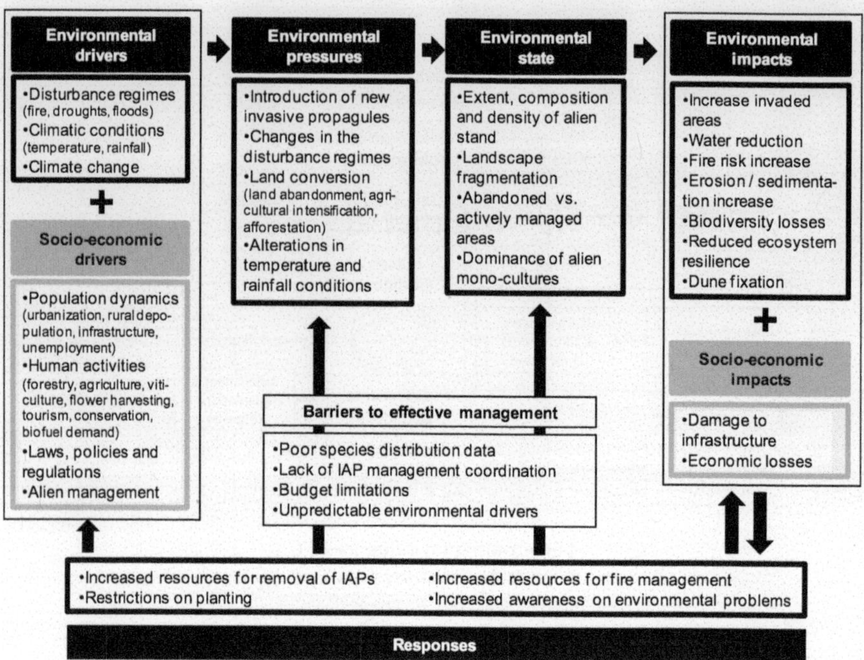

Fig. 1.4 An example of an action-oriented framework: the DPSIR framework for control of invasive plant species. Reproduced from Roura-Pascual et al. (2009) with permission

itself. Bunch et al. (2008) have reviewed seven different but related approaches to management problems in SESs: soft systems methodology (SSM), social systems design, interactive planning, de Beer's viable systems model and team syntegrity, participatory action research, and adaptive management. Walker et al. (2002) propose yet another problem-oriented framework (Fig. 1.5).

The various frameworks and "design principles" developed by Elinor Ostrom and her collaborators (e.g., see Ostrom 1990, 2007, 2009; Crawford and Ostrom 1995; Gibson et al. 2005; Anderies et al. 2004) are difficult to fit into this typology because they have been developed with several different objectives in mind. For example, the Institutional Analysis and Development (IAD) Framework (Crawford and Ostrom 1995; Imperial 1999; reviewed more recently in Gibson et al. 2005) and its successors are problem-oriented frameworks in many ways, because they are focused on solving the problem of developing institutions which facilitate the sustainable use of natural resources, but they have ingredients of hypothesis-testing, action (including "design"), and assessment. They have also contributed usefully to theoretical development through their emphasis on mechanisms, rules, and system description; and they have successfully challenged other, simpler theories such as Hardin's "Tragedy of the Commons" (1968). I have thus included them in the next category, theory-oriented frameworks, despite their problem-oriented facade.

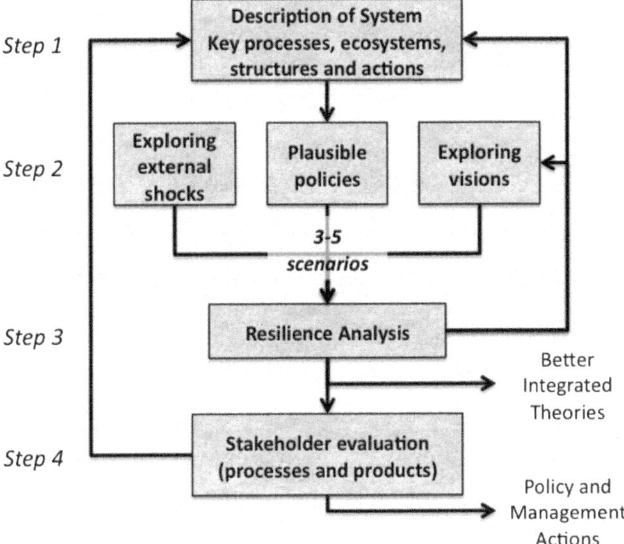

Fig. 1.5 An example of a problem-oriented framework: the resilience analysis framework of Walker et al. (2002). Reproduced with permission from *Ecology & Society*

1.4.5 Theory-Oriented Frameworks

These are the frameworks that attempt to define and connect different pieces of theory within the domain of a particular area of research. There have been a number of attempts at producing general frameworks which either deal directly with SES theory or with relevant aspects of related theories (e.g., see Schwaninger 2006). Understanding these frameworks and building on them is important for the future development of a scientific theory of SESs. Rather than attempting to review all existing frameworks, a book-length project on its own, I will first summarize what I think a general theoretical framework for SESs should aim for, and then highlight some areas in which existing theory-oriented frameworks do a good job of meeting one or more of the criteria. Note that some well-known SES-related frameworks, such as the vulnerability assessment framework and the framework provided by adaptive management, are not included in the next section because they were developed to address management or assessment problems rather than explicitly to frame and advance theory.

1.5 Goals for a Theoretical Framework for SESs

A useful framework for the theoretical development of studies of SESs needs to meet a range of different criteria. These may be both practical and theoretical. The criteria for evaluating a framework could be framed in a number of different ways, depending on the focus of the analyst, and it would be possible to argue about them indefinitely. Despite their almost inevitably controversial nature, however, assessment criteria are unavoidable for any critical comparison. As a basis for comparison and discussion, I therefore propose the following seven criteria for theory-oriented frameworks for the study of SESs:

1. Social-ecological core: a framework may have its origins in either the social or the ecological sciences, but it needs to provide a clear way of linking social and ecological systems and to be strong in both disciplines. Frameworks that deal primarily with economies and claim to be interdisciplinary because they mention ecosystem goods and services, or frameworks created for ecosystems which indirectly include anthropogenic drivers of habitat change, do not fit this criterion. It also excludes conceptual frameworks which offer general ways of thinking about the world, such as integral theory (e.g., Wilber 1997), but do not make specific claims about social-ecological relationships.
2. Empirical support and translation modes: frameworks that claim to be scientific, no matter how elegant, should be supported by rigorous empirical studies. Analyses, results, and conclusions should be framed in a way that is repeatable, at least in principle, and different scientists should ideally reach the same conclusions independently. The criterion of empirical support also includes Popper's falsification criterion; it should be possible in principle to find counter-examples or to disprove empirical claims. Likewise, frameworks should include translation modes that allow theory to be connected to empirical observations, and vice versa. Theory should provide a way of distinguishing between significant and irrelevant observations; and, conversely, observation should provide a way of distinguishing between significant and irrelevant theories. This is not possible if the predictions of a theory cannot be framed in terms of testable hypotheses.
3. Mechanisms: frameworks should offer insights into causality. They should ideally be based on first principles, or at least on accepted observations, and should offer clear statements of cause and effect. Frameworks for SESs should also offer explanations for the complex behaviors observed in real-world SESs. System descriptions alone, whether of system elements or system behaviors, do not provide a complete framework.
4. Spatiotemporal dynamics: frameworks should deal with the dynamic aspects of SESs and the nature of change through time, as well as with the spatial nature of SES and spatial variation.
5. Disciplinary context: frameworks should relate to previous frameworks and, ideally, should be able to explain their weaknesses and/or incorporate their strengths. In a discipline such as physics, for example, the theory of relativity builds on and expands Newtonian physics rather than discarding or ignoring it. In my subjective view, the study of SESs has suffered from an excess of isolated development of frameworks with too little synthesis between frameworks and too much ignorance of preceding ideas.

6. Interdisciplinarity and transdisciplinarity: this builds on the preceding criterion of disciplinary context, but more broadly. Frameworks for SESs should be able to cope with, and offer connections between, complementary perspectives and different epistemologies.
7. Direction: frameworks should provide direction for the study of SESs by suggesting or guiding new empirical studies which will advance our theoretical understanding of SESs.

1.6 Pieces of the Puzzle that Appear in Current Frameworks

Although I do not consider that any existing framework meets all of the seven criteria proposed above, potentially useful theoretical elements appear in many existing theory-oriented frameworks for SES research. The following (and highly subjective) discussion of different frameworks that excel in different assessment criteria is intended to provide a better idea of the kind of overarching synthesis which will be needed to produce a more cohesive, consistent theory of SESs.

1.6.1 Social-Ecological Core

Several existing frameworks provide theoretical guidance to link the analysis of ecosystems and societies. From my perspective as an ecologist, the strongest of these is the cluster of related frameworks around social-ecological resilience and complex systems theory. Panarchy, the Resilience Alliance workbook, and a more recent synthesis of complexity theory from a resilience perspective (A–N–IP; Norberg and Cumming 2008) currently offer some of the clearest expressions of these ideas.

The panarchy framework (Holling 2001; Holling and Gunderson 2002) proposes that social–ecological systems are driven by a series of interconnected adaptive cycles on different scales. The adaptive cycle offers a model of the process of change in a generic SES. The underlying philosophy of the framework is one of continual, nonlinear, episodic change in linked social–ecological systems. Panarchy proposes that complex systems follow adaptive cycles interactively at several different scales (Holling 2001; Holling and Gunderson 2002; Gunderson and Holling 2002). Cycles may be out of synchrony, with phases complementing one another to increase system resilience, or (less commonly, but sometimes catastrophically) in synchrony. The adaptive cycle itself consists of four stages, labeled r, k, Ω (omega), and α (alpha). Adaptive cycles have been observed in societies, economies, and ecosystems, as well as in linked systems, suggesting that panarchy captures something fundamental about the relationships between scale, succession, resource acquisition, and system collapse and regeneration.

The Resilience Alliance workbook (The Resilience Alliance 2007a, b) builds on panarchy but introduces additional independent elements. It focuses on describing and modeling social–ecological systems in the context of sustainable natural resource management. It assumes a multivariate, ever-changing world composed of multiple interacting systems with complex dynamics. The workbook offers an interdisciplinary synthesis which includes elements of ecology, sociology, economics, and complexity theory (including ideas of nonlinearities, regime shifts, and feedbacks). The adaptive cycle (as discussed above) is frequently used as a metaphor for dynamic change but is not essential for applying the ideas presented in the workbook.

The A–N–IP (Asymmetries, Networks, Information Processing) framework is outlined in book form in *Complexity Theory for a Sustainable Future*, edited by Norberg and Cumming (2008). It builds on the frameworks offered by panarchy and the RA workbook, with many of the same contributors. The framework builds on and extends the two previous frameworks (panarchy and the Resilience Alliance workbook) and attempts to place them more clearly within a complexity theory context. It considers SESs through three complementary lenses: asymmetries, networks, and information processing. The framework offers a number of linkages between classical complex systems research and the growing literature on resilience and sustainability. Asymmetries refer to differentiation of components, whether in system elements or along environmental gradients; networks refer to the connections between different system elements, as well as the degree to which they influence and are influenced by membership in a broader system; and information processing focuses on decision making in both an active and a passive sense. The underlying logic of the book is that simultaneously exploring asymmetries, networks, and information processing in a given complex system (or whichever perspective is more appropriate for a given question) will provide a series of different insights which will allow scientists to cope better with, analyze, and understand the different aspects of complexity.

1.6.2 Empirical Support and Translation Modes

Each of the frameworks discussed in this chapter, with the possible exception of the SOHO framework, has some empirical support. There is a considerable amount of empirical support for systems approaches to SES analysis (reviewed in Cumming 2011). Overall, however, the *framework* with, in my opinion, the best balance between empirical observation and theory is probably the sustainability framework of Ostrom (2007, 2009) and collaborators. The sustainability framework focuses on institutions (e.g., rules, laws, customs, and traditions) as a central theme in natural resource management. It is primarily concerned with the question of the ways in which rules and institutions affect the incentives confronting actors. The sustainability framework divides SESs into four basic interacting entities (termed "first-level core subsystems"): resource units, a resource system, a governance system, and users. The framework attempts to "organize how these attributes may affect and be affected by larger socioeconomic, political, and ecological settings in which they are embedded, as well as smaller ones." It thus provides a common set of variables for organizing comparisons of similar SESs, while clarifying a number of expected interactions and system dynamics. The empirical rigor of the sustainability framework probably relates to its origins as a response to the Garret Hardin (1968) "Tragedy of the Commons"; proving that the Tragedy of the Commons was not inevitable required careful assembly of convincing amounts of evidence from within a conceptual framework in which hypotheses and the nature of proof were clearly and explicitly stated.

1.6.3 Mechanisms

Keeping in mind that mechanistic frameworks are ideally based on first principles, I am only aware of two existing frameworks that meet this particular criterion well. The first is the thermodynamic framework ("SOHO") of Kay and Boyle (2008). The second is that of agent-based models.

The SOHO framework (Kay and Boyle 2008) suggests a physical basis for self-organization as a consequence of the dissipation of energy. Exergy refers to the quality of energy, particularly its ability to do useful work. The proponents of the framework argue that as systems move further from equilibrium, exergy increases, more dissipative opportunities become available, and more organization emerges. Flows from ecosystems provide exergy both supporting and constraining human society. The SOHO framework assumes hierarchies of structures and processes and focuses on flows, feedbacks, and thresholds. Although SOHO offers a potentially appealing physical basis for the emergence of complexity, it sheds little light on how and why particular kinds of structure emerge in complex systems or why they behave as they do. To draw a parallel to a purely ecological theory, although Elton's trophic pyramid applies the laws of thermodynamics to explain why the numbers or biomass of organisms at each trophic level differ predictably, it does not provide insights into the emergence of particular aspects of the food web structure. In the case of SESs, the finer details are often what we are interested in, rather than the coarser reasons for the emergence of complexity; the SOHO framework thus ends up a level of detail short in terms of its relevance to the analysis of SESs.

Agent-based models offer another mechanistic alternative. The primary philosophy of agent-based models is one of "bottom-up" simulation of complex adaptive systems (Grimm et al. 2005). Several different frameworks have been proposed for working with multi-agent models. Holland (1992, 1995) has perhaps come closest to developing a general framework by laying out seven key elements of complex systems: four properties (aggregation, nonlinearity, flows, and diversity) and three mechanisms (tags, internal models, and building blocks). Holland (1995) argued that these elements represent the minimal set of properties and mechanisms from which real-world complexity emerges. He assembled them in a hierarchical (two-tiered) arrangement to construct an approach ("Echo", a multi-agent model) to simulating complex adaptive systems. The basic elements of most current multi-agent models are individuals (Grimm et al. 2005). As in Holland's outline, these individuals, or agents, are usually characterized by possessing internal states (goals and representations of the environment and other agents) and the abilities to communicate, to act, and to perceive. They exist in an external environment, which may be heterogeneous and is usually spatially explicit (i.e., each agent has a location in space and time, and interactions between agents may be limited to occurring when they are in close proximity to one another).

1.6.4 Spatiotemporal Dynamics

Many of the previous frameworks, with the obvious exception of multi-agent models, have been developed from a basis that is not spatially explicit. Two of the current crop of frameworks focus explicitly on spatial variation. These are the "Three Critical Dimensions of Land Use and Land Cover Change (3CD)" framework (Agarwal et al. 2002) and the Spatial Resilience framework (Cumming 2011).

The 3CD framework was developed by Agarwal et al. (2002) as part of a review of LULCC models. The framework is explicitly hierarchical in space and time. It was developed to compare different LULCC models, including explicit consideration of model scale (both grain and extent) and model complexity. Agarwal et al. (2002) further distinguish (along the human decision-making axis) between agent and domain. Agent "refers to the human actor or actors in the model who are making decisions," while domain "refers to the broadest social organization incorporated in the model." Agents make decisions, while domains define institutional and geographic contexts. This framework thus contains an obvious connection to agent-based modeling.

The spatial resilience framework (Cumming 2011) builds on the frameworks of panarchy, the resilience alliance workbook, and A–N–IP discussed above. Spatial resilience has elements which are both internal and external to the system. The primary internal elements of spatial resilience include: the spatial arrangement of system components and interactions; spatially relevant system properties, such as system size, shape, and the number and nature of system boundaries (e.g., hard or soft, and whether temporally variable or fixed over time scales of interest); spatial variation in internal phases, such as vegetation successional stage, which influence resilience; and unique system properties that are a function of location in space. The primary external elements of spatial resilience include: context (spatial surroundings, defined at the scale of analysis); connectivity (including spatial compartmentalization or modularity); and resulting spatial dynamics, such as spatially driven feedbacks and spatial subsidies. Both internal and external elements must be considered in relation to other aspects of system resilience, including such things as the number and nature of components and interactions, the ability of the system to undergo change while maintaining its identity, system memory, and the potential inherent in the system for adaptation and learning. Spatial resilience thus uses location and spatial relationships to connect different system elements and provide an integrated platform for the analysis of social–ecological systems.

1.6.5 Disciplinary Context

The majority of published frameworks for research on SESs have done a relatively poor job of expanding their breadth and reviewing or considering other frameworks that have attempted to address the same questions. This is one of the central problems in the field. In my subjective view, the study of SESs has suffered from the isolated development of frameworks, with too little synthesis between frameworks and too much disregard for preceding ideas. Alternative frameworks are often perceived as competing memes, rather than as complementary approaches, and each group that battles with the problem seems to resolve it in a different way. For example, most authors writing on resilience and adaptive management have ignored the relevant insights provided by Soft Systems Analysis (Checkland 1981; see also Cundill et al. 2012).

1.6.6 Interdisciplinarity and Transdisciplinarity

Despite the recent rise of interdisciplinary science, one of the weakest links in the study of social–ecological systems remains the one between different epistemologies. Ecologists, economists, and sociologists, for example, often appear to be speaking different languages and frequently reject suggestions or approaches developed in other disciplines. This disciplinary snobbishness, coupled with the ignorance that usually goes with it, remains a major barrier to advances in the development of general theories of SESs.

Communicating ideas about SESs to stakeholders and managers appears to work well in many cases, and in some instances—such as the development of management strategies for Kruger National Park in South Africa—one or more theoretical frameworks have been effectively used to bridge transdisciplinary gaps. In many cases, however, transdisciplinarity is achieved more by other kinds of framework that are loosely derived from a theoretical framework but differing in key aspects. For example, adaptive management has become a widely used approach to natural resource management, but its adoption does not automatically incorporate into management the set of important ideas about thresholds, adaptive cycles, and the relevance of nonlinear processes that are associated with panarchy.

1.6.7 Direction

This criterion requires the framework to highlight areas in which current knowledge is deficient and new investigations are needed. Nearly all of the frameworks I have discussed here, and many others I have not, offer suggestions for new research directions (e.g., see the propositions outlined by Walker et al. 2006). The greater challenge is to synthesize theory in such a way as to highlight directions which will be fruitful and distinguish them from those less relevant. One of the most important considerations in highlighting new directions for research is that they should be concerned with developing broadly useful generalities, rather than refining or testing specific approaches or concepts.

1.7 Discussion

When considering the match between assessment criteria and existing frameworks, it quickly becomes apparent that no existing framework meets all the requirements for a cohesive theory of SESs. Each of the eight individual frameworks discussed in this text has its own strengths and weaknesses. In addition, different people have different perceptions of what each framework entails and of its relative value as an organizing framework for research on social–ecological systems.

It is interesting that no single framework is dominant across all the different areas to which a framework could be expected to contribute. Most existing frameworks appear to have been developed from a particular starting point and to have remained strong in their original area but not to have crossed conceptual dividing lines significantly. For example, multi-agent models offer a good way to explore and interpret social-ecological complexity, but the framework remains relatively weak on translation modes and is somewhat lacking in empirical support—a clear challenge to members of the multi-agent modeling community to develop a stronger case for the real-world relevance of their approach. The tendency of researchers to develop "new" frameworks without fully explaining how they relate to other existing frameworks and what new elements they bring to the problem is another obvious reason for the lack of a single dominant, unifying framework.

One of the tensions in many case studies of social–ecological systems appears to be between documentation and explanation; description of the different components and the dynamics of a real-world SES can be time-consuming and technically challenging, and frequently becomes the dominant theme in a framework. By contrast, frameworks focusing on first principles and mechanisms creating complexity are often difficult to apply to the real world and insufficiently grounded in empirical data. This tension is nicely illustrated in the excellent book by Waltner-Toews et al. (2008), which includes as part of "the ecosystem approach" both a theory-oriented framework (SOHO, discussed above) and a problem-oriented framework, termed AMESH (Adaptive Methodology for Ecosystem Sustainability and Health). AMESH offers a useful problem-oriented framework for case study analysis but, as presented in the book, it appears unrelated to the SOHO framework. From a practical perspective, AMESH is far more useful; while from a theoretical perspective, SOHO provides a potential set of unifying concepts but no obvious way to move from theory to application. A theoretical framework should ideally be able to explain a real-world phenomenon, in the same way that Newton's laws explain the motions of the planets, to a relatively high level of detail and precision.

Consideration of the theoretical basis for the analysis of SESs also highlights the many commonalities between different approaches to studying social–ecological systems. Many of the central concerns of SES theory are similar to those arising in other areas of science. They include, for example, understanding the relationships between structure and function, and between pattern and process; the degree to which pattern-process or structure-function relationships translate across or between scales; the relative importance of internal and external variation as drivers of system dynamics; and establishing the degree to which interpolation and prediction using standard scientific methods can be successful. These questions, and others like them, suggest an important general set of targets for further theoretical development and empirical investigation in the study of SESs.

SES theory also has some additional concerns that differ from the central themes of classical science. These additions include such things as the need to understand better the research process itself; concerns over the impact of the phrasing and delineation of research questions on the eventual conclusions of the study; a necessary preoccupation with self-organization, thresholds, and nonlinearities; and a continual need to think about variables which are highly unpredictable and potentially difficult to quantify, such as innovations, perturbations, and learning. These elements of SES theory suggest that theoretical frameworks for the analysis of SESs will never be able to rely purely on systems of equations or descriptions of physical phenomena. Rather, because of the "Social" in SES, they will need to take into account the unique properties of social systems and the unavoidable subjectivity involved in analyzing ourselves. Developing a theoretical framework for SESs will thus require a truly transdisciplinary synthesis of ideas and concepts and quite possibly an outline that is composed of several different modules in a hierarchical arrangement (e.g., Cumming 2011). Although the frameworks discussed in this chapter offer a solid starting point, there is still a long way to go before we can claim to have a rigorous theoretical framework for the analysis of social–ecological systems.

References

Adger WN (2006) Vulnerability. Glob Environ Change 16:268–281

Adger WN, Jordan A (2009) Sustainability: exploring the processes and outcomes of governance. In: Adger WN, Jordan A (eds) Governing sustainability. Cambridge University Press, Cambridge, pp 3–31

Adger WN, Hughes T, Folke C, Carpenter SR, Rockström J (2005) Social-ecological resilience to coastal disasters. Science 309:1036–1039
Agarwal C, Green GM, Grove J, Evans TP, Schweik CM (2002) A review and assessment of land-use change models: dynamics of space, time, and human choice. In: General Technical Report NE-297. US Department of Agriculture Forest Service, Northeastern Research Station
Agrawal A, Gibson CC (1999) Enchantment and disenchantment: the role of community in natural resource conservation. World Dev 27:629–649
Anderies JM, Janssen MA, Ostrom E (2004) A framework to analyze the robustness of social-ecological systems from an institutional perspective. Ecol Soc 9(1):18
Berkes F, Colding J, Folke C (2003) Navigating social-ecological systems: building resilience for complexity and change. Cambridge University Press, Cambridge
Bunch M, McCarthy D, Waltner-Toews D (2008) A family of origin for an ecosystem approach to managing for sustainability. In: Waltner-Toews D, Kay JJ, Lister N-ME (eds) The ecosystem approach: complexity, uncertainty, and managing for sustainability. Columbia University Press, New York, pp 125–138
Carpenter SR (2002) Ecological futures: building an ecology of the long now. Ecology 83:2069–2083
Checkland P (1981) Systems thinking, systems practice. Wiley, Chichester
Connell JH (1978) Diversity in tropical rain forests and coral reefs. Science 199:1302–1310
Costanza R, Andrade F, Antunes P, van den Belt M, Boesch D, Boersma D, Catarino F, Hannah S, Limburg K, Lowj B, Molitor M, Pereira JG, Rayner S, Santos R, Wilson J, Young M (1999) Ecological economics and sustainable governance of the oceans. Ecol Econ 31:171–187
Crawford S, Ostrom E (1995) A grammar of institutions. Am Pol Sci Rev 89:582–600
Cumming GS (2011) Spatial resilience in social-ecological systems. Springer, Berlin
Cundill G, Cumming GS, Biggs D, Fabricius C (2012) Soft systems thinking and social learning for adaptive management. Conserv Biol 26:13–20
Folke C (2006) Resilience: the emergence of a perspective for social-ecological systems analyses. Glob Environ Change 16:253–267
Folke C, Holling CS, Perrings C (1996) Biological diversity, ecosystems, and the human scale. Ecol Appl 6:1018–1024
Gergel SE, Bennett EM, Greenfield BK, King S, Overdevest CA, Stumborg B (2004) A test of the environmental Kuznets curve using long-term watershed inputs. Ecol Appl 14:555–570
Gibson CC, Andersson K, Ostrom E, Shivakumar S (2005) The Samaritan's dilemma: the political economy of development aid. Oxford University Press, Oxford
Glymour C, Danks D, Glymour B, Eberhardt F, Ramsey J, Scheines R, Spirtes P, Teng CM, Zhang J (2009) Actual causation: a stone soup essay. Synthese. doi:10.1007/s11229-009-9497-9
Grimm V, Revilla E, Berger U, Jeltsch F, Mooij WM, Railsback SF, Thulke H-H, Weiner J, Weigand T, DeAngelis D (2005) Pattern-oriented modeling of agent-based complex systems: lessons from ecology. Science 310:987–991
Gunderson LH, Holling CS (eds) (2002) Panarchy: understanding transformations in human and natural systems. Island Press, Washington
Hardin G (1968) The tragedy of the commons. Science 162:1243–1248
Holland JH (1992) Adaptation in natural and artificial systems: an introductory analysis with applications to biology, control, and artificial intelligence, 2nd edn. MIT Press, Cambridge
Holland JH (1995) Hidden order: how adaptation builds complexity. Perseus Books, New York
Holling CS (2001) Understanding the complexity of economic, social and ecological systems. Ecosystems 4:390–405
Holling CS, Gunderson LH (2002) Resilience and adaptive cycles. In: Gunderson LH, Holling CS (eds) Panarchy: understanding transformations in human and natural systems. Island Press, Washington, pp 25–62
Holling CS, Meffe GK (1996) Command and control and the pathology of natural resource management. Conserv Biol 10:328–337

Imperial MT (1999) Institutional analysis and ecosystem-based management: the institutional analysis and development framework. Environ Manag 24:449–465

IPCC (2007a) Climate change (2007): impacts, adaptation and vulnerability. In: Contribution of working group II to the fourth assessment report of the intergovernmental panel on climate change. Cambridge University Press, Cambridge

IPCC (2007b) Climate change (2007): the physical science basis. In: Contribution of working group I to the fourth assessment report of the intergovernmental panel on climate change. Cambridge University Press, Cambridge

Johns D (2009) A new conservation politics: power, organization building and effectiveness. Wiley, Chichester

Kay J, Boyle M (2008) Self-organizing, holarchic, open systems (SOHOs). In: Waltner-Toews D, Kay JJ, Lister N-M (eds) The ecosystem approach: complexity, uncertainty, and managing for sustainability. Columbia University Press, New York

Koontz TM, Bodine J (2008) Implementing ecosystem management in public agencies: lessons from the U.S. Bureau of land management and the forest service. Conserv Biol 22:60–69

Kuhn T (1962) The structure of scientific revolutions. University of Chicago Press, Chicago

Kuznets S (1955) Economic growth and income inequality. Am Econ Rev 45:1–28

Millennium Assessment (2003) Ecosystems and human well-being: a framework for assessment. Island Press, Washington

Millennium Assessment (2005) Ecosystems and human well-being: biodiversity synthesis. Island Press, New York

Norberg J, Cumming GS (2008) Complexity theory for a sustainable future. Columbia University Press, New York

Ostrom E (1990) Governing the commons: the evolution of institutions for collective action. Cambridge University Press, Cambridge

Ostrom E (2007) A diagnostic approach for going beyond panaceas. Proc Natl Acad Sci 104:15181–15187

Ostrom E (2009) A general framework for analyzing sustainability of social-ecological systems. Science 352:419–422

Pickett STA, Jones C, Kolasa J (2007) Ecological understanding: the nature of theory and the theory of nature. Academic, New York

Robbins P (2004) Political ecology: a critical introduction. Blackwell Publishing, Malden

Roura-Pascual N, Richardson DM, Krug RM, Brown A, Chapman RA, Forsyth GG, Le Maitre DC, Robertson MP, Stafford L, Van Wilgen BW, Wannenburgh A, Wessels N (2009) Ecology and management of alien plant invasions in South African fynbos: accommodating key complexities in objective decision making. Biol Conserv 142:1595–1604

Schwaninger M (2006) System dynamics and the evolution of the systems movement. Syst Res Behav Sci 23:583–594

Stommel H (1963) The varieties of oceanographic experience. Science 139:572–576

The Nature Conservancy (TNC) (2003) The five-S framework for site conservation: a practitioner's handbook for site conservation planning and measuring conservation success, 3rd edn. http://www.nature.org

The Resilience Alliance (2007a) Assessing and managing resilience in social-ecological systems: a practitioner's workbook, vol 1. version 1.0. http://www.resalliance.org/3871.php

The Resilience Alliance (2007b) Assessing resilience in social-ecological systems: a scientist's workbook. http://www.resalliance.org/3871.php

Turner MG, Gardner RH, O'Neill RV (2001) Landscape ecology in theory and practice: pattern and process. Springer, New York

Turner BLI, Kasperson RE, Matson PA, McCarthy JJ, Corell RW, Christensen L, Eckley N, Kasperson JX, Luers A, Martello ML, Polsky C, Pulsipher A, Schiller A (2003a) A framework for vulnerability analysis in sustainability science. Proc Natl Acad Sci 100:8074–8079

Turner BLI, Matson PA, McCarthy JJ, Corell RW, Christensen L, Eckley N, Hovelsrud-Broda GK, Kasperson JX, Kasperson RE, Luers A, Martello ML, Mathiesen S, Naylor R, Polsky C,

Pulsipher A, Schiller A, Selin H, Tyler N (2003b) Illustrating the coupled human-environment system for vulnerability analysis: three case studies. Proc Natl Acad Sci 100:8080–8085

Walker B, Carpenter S, Anderies J, Abel N, Cumming GS, Janssen M, Lebel L, Norberg J, Peterson GD, Pritchard R (2002) Resilience management in social-ecological systems: a working hypothesis for a participatory approach. Conserv Ecol 6:14

Walker BH, Gunderson LH, Kinzig AP, Folke C, Carpenter SR, Schultz L (2006) A handful of heuristics and some propositions for understanding resilience in social-ecological systems. Ecol Soc 11:13

Waltner-Toews D, Kay JJ, Lister N-ME (eds) (2008) The ecosystem approach: complexity, uncertainty, and managing for sustainability. Columbia University Press, New York

Wilber K (1997) An integral theory of consciousness. J Conscious Stud 4:71–72

Part II
Ecosystems Under Human Activities

Chapter 2
The Extent of Biodiversity Recovery During Reforestation After Swidden Cultivation and the Impacts of Land-Use Changes on the Biodiversity of a Tropical Rainforest Region in Borneo

Kohei Takenaka Takano, Michiko Nakagawa, Takao Itioka, Keiko Kishimoto-Yamada, Satoshi Yamashita, Hiroshi O. Tanaka, Daisuke Fukuda, Hidetoshi Nagamasu, Masahiro Ichikawa, Yumi Kato, Kuniyasu Momose, Tohru Nakashizuka, and Shoko Sakai

K.T. Takano (✉)
Research Institute for Humanity and Nature, Kyoto, Japan

Graduate School of Life Sciences, Tohoku University, Sendai, Japan
e-mail: kohei@m.tohoku.ac.jp

M. Nakagawa
Graduate School of Bioagricultural Sciences, Nagoya University, Nagoya, Japan

T. Itioka
Graduate School of Human and Environmental Studies, Kyoto University, Kyoto, Japan

K. Kishimoto-Yamada
Graduate School of Arts and Sciences, The University of Tokyo, Tokyo, Japan

S. Yamashita
Forestry and Forest Products Research Institute, Tsukuba, Japan

H.O. Tanaka • T. Nakashizuka
Graduate School of Life Sciences, Tohoku University, Sendai, Japan

D. Fukuda
Center for Ecological Research, Kyoto University, Otsu, Japan

H. Nagamasu
The Kyoto University Museum, Kyoto, Japan

M. Ichikawa
Kochi University, Kochi, Japan

Y. Kato
Hakubi Center for Advanced Research, Kyoto University, Kyoto, Japan

K. Momose
Department of Forest Resources, Ehime University, Matsuyama, Japan

S. Sakai
Research Institute for Humanity and Nature, Kyoto, Japan

Center for Ecological Research, Kyoto University, Otsu, Japan

S. Sakai and C. Umetsu (eds.), *Social–Ecological Systems in Transition*, Global Environmental Studies, DOI 10.1007/978-4-431-54910-9_2, © Springer Japan 2014

Abstract Whereas many studies have addressed the effects of deforestation on biodiversity, few have focused on the recovery of diversity during reforestation. This study aimed at evaluating the recovery of, or chronosequential changes in, the biodiversity in the fallows (i.e., secondary vegetation or forests that form during the resting periods following harvest in shifting cultivation) of a tropical rainforest region in Borneo. We also aimed at determining the impacts of forest fragmentation and other land-use changes on biodiversity.

We established several study plots in fallows at different stages of succession, specifically, new fallows (rested for 1–3 years), young fallows (rested for 5–13 years), and old fallows (rested for 20–60 years). We also established study plots in a continuous primary forest and fragmented primary forests, extensive rubber gardens, and other land-use types. In addition, we investigated the diversity of trees, fungi, and animals, and compared the values obtained among the different land-use types.

With each progression in forest stage, the species richness, species density, and encounter rates increased for trees, fungi, army ants, and ants attending hemipterans. However, even after fallow periods of 20–60 years, the levels of diversity had not fully recovered to those recorded in primary forests. The biodiversity indices of fragmented primary forests were lower compared to those of the continuous primary forest for army ants. The biodiversity indices of extensive rubber gardens and the other land-use types were also lower compared to those of the continuous primary forest for bats. Such trends were also observed for aphyllophoraceous fungi and some insects.

These results indicate the irreplaceable value of continuous primary forests for conserving biodiversity. In contrast, the species richness of small mammals and phytophagous scarabaeid beetles was similar or even higher in fragmented primary forests and the other land-use types compared to the continuous primary forest. Further studies are necessary to investigate how the characteristics of each taxonomic group (e.g., disturbance tolerance, dispersal ability, and the life history traits) are related to the different types of disturbance (intensity, spatio-temporal configurations, and the consequent changes in the environmental factors of each habitat type).

Keywords Biodiversity • Fallow and secondary forest • Lambir Hills National Park • Sarawak • Swidden, shifting, and slash-and-burn agriculture

2.1 Introduction

A number of parameters contributing to the loss of biodiversity in tropical forests have been extensively studied, including the effects of forest fragmentation (e.g., Debinski and Holt 2000; Tscharntke et al. 2007), conventional and reduced-impact logging (RIL: e.g., Samejima et al. 2012), and land-use changes from primary forests to secondary forests or plantations (e.g., Barlow et al. 2007).

However, very few studies have focused on the recovery of biodiversity as forests regenerate to form old secondary forests (but see, e.g., Veddeler et al. 2005).

For centuries, swidden cultivation (i.e., shifting or slash-and-burn cultivation) has been an integral part of landscapes in tropical forests, which are crucial for biodiversity conservation (Mertz et al. 2009b; Padoch and Pinedo-Vasquez 2010). Some scholars regard swidden cultivation as a primary cause of deforestation and forest degradation (Freeman 1955; Lanly 1982). In contrast, other scholars claim that some traditional swidden cultivation is a sustainable practice for maintaining ecosystem services because of the complex patterns of fallow mosaics, which contain a mixture of vegetation at various stages of regrowth (Rerkasem et al. 2009; Padoch and Pinedo-Vasquez 2010).

To devise better strategies for biodiversity conservation in tropical forest regions, it is important to determine (1) the extent to which biodiversity recovers during the reforestation process in fallow periods of swidden cultivation and (2) the impacts of land-use changes and forest fragmentation on biodiversity.

2.1.1 The Tropical Rainforest of the Lambir Hills National Park (LHNP)

The primary lowland forests of the Southeast Asian tropics are characterized by extremely species-rich biodiversity (Whitmore 1998). The current project was conducted in and around the Lambir Hills National Park (LHNP; 4°08–12′ N, 114°00′–07′ E, 20–465 m a.s.l.), which is situated approximately 20 km south of Miri City and 10 km from the northwest coast of Borneo, in the northern part of the state of Sarawak, Malaysia (Fig. 2.1). The LHNP was established in 1975, and forms part of a forest reserve in a water catchment area for Miri City, covering 6,949 ha of land (Hazebroek and Abang Kashim bin Abang Morshidi 2000; Ichikawa 2007). The average annual rainfall and mean air temperature in the LHNP from 2000 to 2009 were 2,600 mm and 25.8 °C, respectively (Kume et al. 2011). The region is subject to irregular and short-term droughts; however, there is no regular dry season, as such (Kume et al. 2011). The national park is mainly covered with a primary lowland mixed dipterocarp forest (Hazebroek and Abang Kashim bin Abang Morshidi 2000), which is formed on nutrient-poor sandy or clay soil (Watson 1985). The forest is multi-layered, with a canopy approximately 35–50 m in height and emergent trees (i.e., trees rising above the canopy) reaching a maximum height of 70 m (Kato et al. 1995). The forest potentially contains the richest tree diversity in the Old World, with a record number of 1,175 tree species in a 52-ha plot (Condit et al. 2000; Ashton 2005; Corlett 2009). Arthropod assemblages are also highly diverse. For example, 347 butterfly species have been reported to occur in the primary forest and surrounding areas of the LHNP (Itioka et al. 2009).

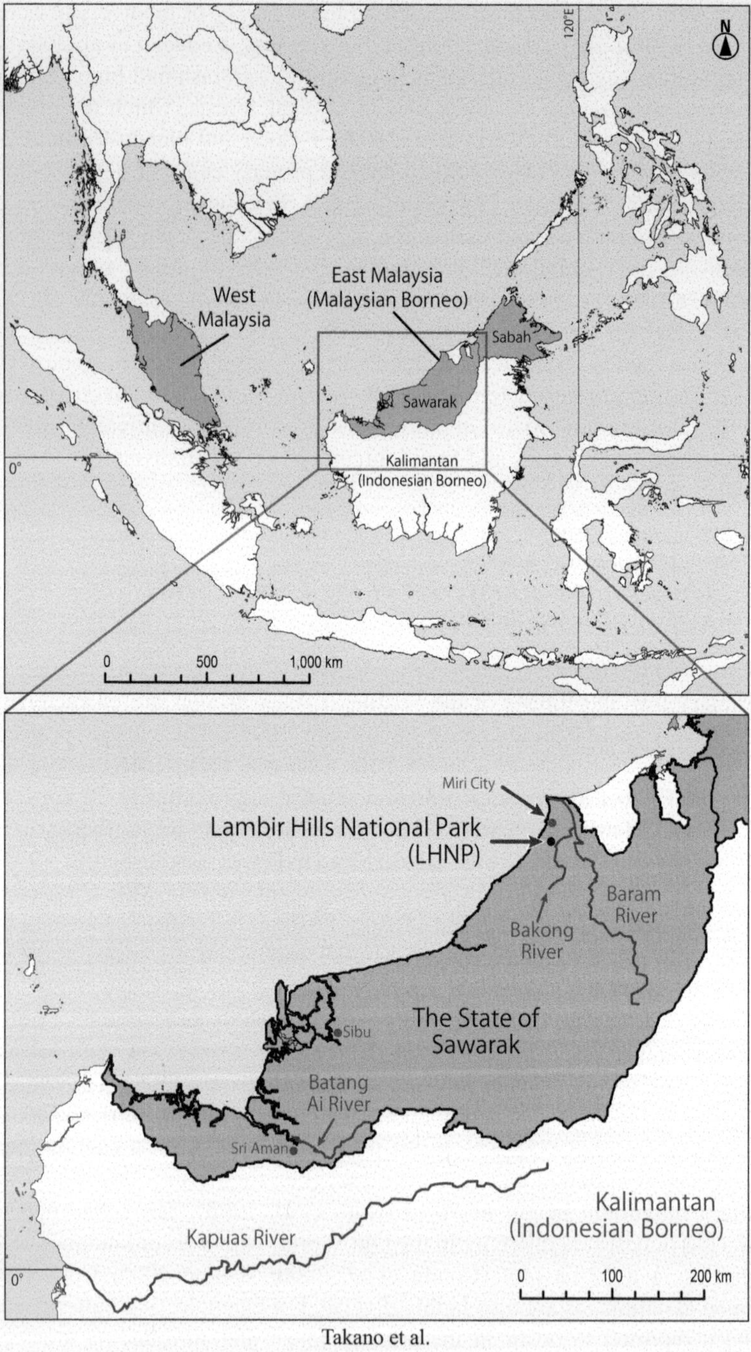

Fig. 2.1 The location of the Lambir Hills National Park in Sarawak, Malaysian Borneo

2.1.2 Traditional Swidden Cultivation by the Iban Around the LHNP

Swidden cultivation in Malaysian Borneo is mainly practiced by more than 50 ethnic groups that have traditionally inhabited rural areas, including the Iban, Bidayuh, Kayan, Murut, and Dusun tribes (Mertz et al. 2009a). Ichikawa (2010) described the practice of swidden cultivation by the Iban who inhabit villages near the LHNP in 1995. In the author's description, forest stands (0.46 ha on average, $n = 14$) of fallows that had rested for periods of 7–10 years were cleared with bush-knives, axes, and fire. Rice and other crops (e.g., maize, cassava, and vegetables) were cultivated for just 1 year. Subsequently, the fields were allowed to revert into secondary forests.

Freeman (1955) reported that, after the clearance of a primary forest stand, the Iban of Sarawak continue cropping for a few years until the productivity of the land declines. In contrast, when a secondary forest stand is cleared, the area is only cropped for 1 year (but see Sect. 2.4.1 of this chapter). A wide range in fallow lengths has been reported, lasting from 3 to more than 25 years (Freeman 1955; Padoch 1982; Cramb 1993). Since the speed of forest succession differs at each location, Iban farmers usually judge where to open a new swidden based on the girth of trees and other forest structures, rather than the duration of fallow periods (Padoch 1982; Dove 1985).

2.1.3 Indigenous Categories of Land-Use by the Iban

The Iban have several land-use categories, specifically: (1) *kampong*, which is a primary forest that has not been farmed within historical memory, including a previously farmed forest that is currently difficult to distinguish from primary forests; (2) *umai*, which is a currently cultivated swidden field; (3) *jeremi* or *redas*, which is a recently fallowed field that has rested for 1 year following the last harvest; (4) *temuda*, which is a young secondary forest with a fallow period of 2–10 years; (5) *damun*, which also refers to a young secondary forest that regenerated during fallow periods of 10–20 years, depending on the nature of the soil and human population pressure; (6) *pengerang, which is an old secondary forest, where* the vegetation resembles that of a primary forest, having been left uncultivated for 20 years or more; (7) *pulau galau*, which is a special tree reserve designated to ensure the availability of logs for house construction; and (8) *kebun*, which contains rubber and fruit tree groves (Table 2.1, Fig. 2.2; Freeman 1955; Wadley et al. 1997; Ichikawa 2007).

Under customary land-use by the Iban, a household that clears a stand of a primary forest gains ownership of the land, and the title of the land is inherited by the family's descendants (Ngidang 2005). However, the Sarawak Land Code limits the right to land first-cut to before 1958 (Ngidang 2005). Therefore, current

Table 2.1 The classification of the study plots in relation to the indigenous categories of land-use by the Iban

Classification of plots in the current study	Continuous primary forest	Not studied	New fallow	Young fallow	Old fallow	Fragmented primary forest	Extensive rubber garden	Fruit tree grove
Explanation	The primary forest in the national park	Currently cultivated swidden fields	Recently fallowed fields rested for 1–3 years after cropping	Younger secondary forests regenerated during fallow periods of 5–13 years	Older secondary forests regenerated during fallow periods of 20–60 years	Special tree reserves for logs for house construction	Small-scale extensive rubber gardens	Fruit tree groves
Indigenous category	*kampong*	*umai*	*jeremi* or *redas*	*temuda* *damun*	*pengerang*	*pulau galau*	*kebun*	
Years after swidden cultivation	No significant disturbance within historical memory	0 year	1 year	2–10 years 10–20 years	≥20 years	No significant disturbance within historical memory	Variable	
Swidden cycle		⋯→	→	→ →	→			
Customary rights	Open access	Individual (household) tenure						

Fig. 2.2 General view of the study plots at different stages of regeneration or subject to different types of land use. (**a**) Currently cultivated swidden field (dissemination after slash and burn). (**b**) New fallow. (**c**) Young fallow. (**d**) Old fallow. (**e**) Continuous primary forest of the Lambir Hills National Park. (**f**) Fragmented primary forest. (**g**) Extensive rubber garden. See Table 2.1 for details about each land-use type. Photographs by Kato Y. (**a**) and Tanaka H.O. (**b–g**)

swidden cultivation is basically repeated within secondary forests, where the customary rights were established before 1958 (but see Ichikawa 2010).

2.1.4 History of Land-Use Alterations in the Study Area

Ichikawa (2003, 2007, 2010) documented the history of land-use change around the LHNP, which is summarized in the following text.

2.1.4.1 Immigration of the Iban

Sarawak is home to approximately 2.7 million inhabitants (Department of Statistics Malaysia 2004) and has a population encompassing various ethnic groups. The largest native ethnic group is the Iban, who represent more than 30 % of the population (Department of Statistics Malaysia 2004). The Iban are believed to have emigrated from the Kapuas River basin, currently West Kalimantan, Indonesia, and settled in Sarawak, on the Malaysian side of Borneo. Sandin (1967) reported that the Iban settled in the Batang Ai River basin, which is close to today's Sri Aman and Sibu Divisions, 14 generations ago. The Iban then emigrated from Sri Aman and Sibu Divisions and settled in the Bakong River basin, which is where the LHNP is located, from the end of the ninetieth century to the early twentieth century (Pringle 1970; Sandin 1994). Before the Iban settled in this region, much of the Bakong area was the territory of hunter-gatherers and was covered with primary forests (Ichikawa 2007).

2.1.4.2 Swidden Cultivation

The immigrants collected forest products to sell, such as wild rubber and rattan, while cutting the primary forest around their longhouses for swidden cultivation. Until the 1960s, hill swiddens were important for the Iban and were larger in size compared to paddy fields established in the swamps. This is because hill rice tasted and smelled better than wet rice at that time, and because other products from hill swiddens were important for subsistence under an economy that was more self-sufficient compared to that of today. The area covered by hill swiddens decreased from 1963 to 1997 (Ichikawa 2007), with almost no swidden cultivation being observed around the LHNP after 2000 (Ichikawa, personal observation). The reasons for the reduction in hill swiddens include (1) an increase in rice production from paddy fields in the swamps after the 1970s and (2) a decrease in the importance of hill swidden products, since it became possible to buy vegetables and root crops from nearby markets(Ichikawa 2007).

2.1.4.3 Other Types of Land Uses

Small-scale patches of Para rubber (*Hevea brasiliensis*, Euphorbiaceae) were planted after swidden cultivation in the 1950s and early 1960s (Ichikawa 2003). Rubber was planted when the international market prices were high; however, the trees were not adequately harvested and managed when rubber prices dropped. Various tree species grow naturally in rubber gardens where rubber tapping has not been conducted for several years.

Commercial selective logging on a relatively small scale was conducted by small Chinese companies after the mid-1970s. Immediately after areas were subjected to logging activity, the Iban cut and burned the remaining trees to create swiddens. The Chinese companies continued to operate until the end of the 1980s, cutting all remaining large-trunked trees.

The Iban began to sell fruit in Miri City after the 1980s. Groves containing various types of fruit trees have been created around work huts that are primarily used for the preparation of rice. From the late 1980s onward, fruit trees were planted around the huts and in any location where secondary forests had been opened.

After 2000, several patches of oil palm gardens (1–2 ha in size) were created near roads to sell the products to nearby oil processing factories at reasonable prices. The oil palm gardens were cultivated by some Iban who had a certain amount of capital and land near roads, from which the products could be easily transported out. In Malaysia, the oil palm area has increased rapidly from 1.8 million ha in 1990 to 4.2 million ha in 2005 (FAO 2010), during which time 1.1 million ha of primary forest have been lost (Fitzherbert et al. 2008).

Consequently, primary forests outside the LHNP had almost completely disappeared by 1997 (97 % of the study area of Ichikawa 2007). Currently, the LHNP is surrounded by a mosaic of fallows, fragmented primary forests, extensive rubber gardens, fruit tree groves, small- to large-scale oil palm plantations, logged forests, and wet paddies.

2.2 Study Design

In total, 46 study plots were established in and around LHNP between 2003 and 2011. The area of each plot ranged in size from approximately 0.1 ha to 0.5 ha (Fig. 2.3). We interviewed land-holders and classified the plots into the following categories: (1) new fallows, rested for 1–3 years after cropping; (2) young fallows, rested for 5–13 years; (3) old fallows, rested for 20–60 years; (4) fragmented primary forests, which have not been significantly disturbed for 100 years or more; and (5) continuous primary forest, which is the protected tropical rainforest in the national park. We also established study plots in extensive rubber gardens, fruit tree groves, and oil palm plantations (Table 2.1, Fig. 2.2).

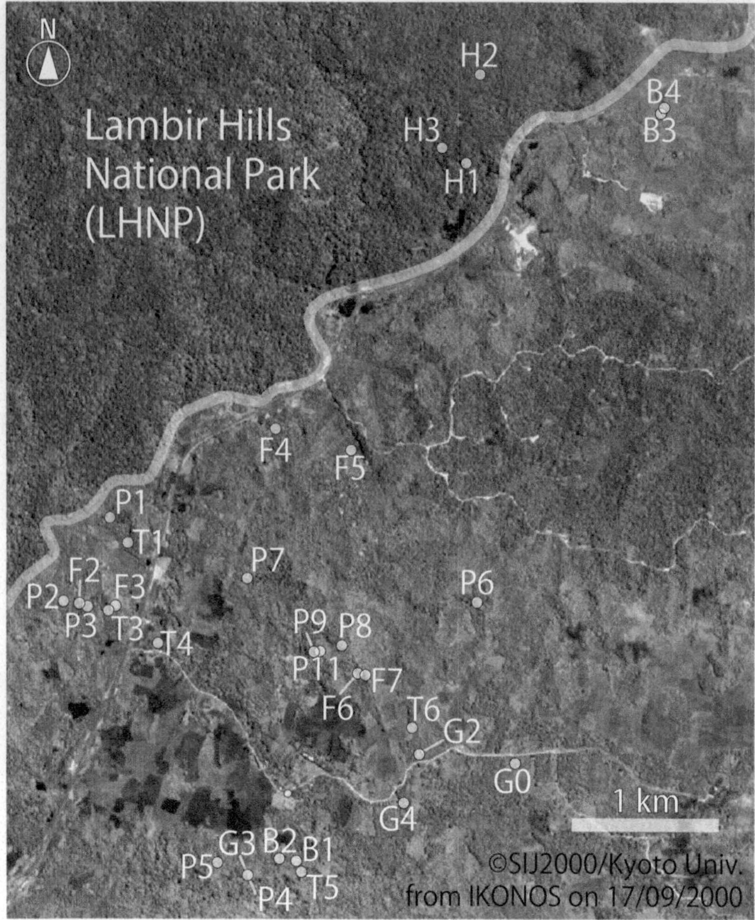

Fig. 2.3 Spatial distribution of the study plots in and around the Lambir Hills National Park (LHNP). Different letters indicate different types of land use: (H) continuous primary forest in the LHNP, (P) fragmented primary forest, (F) old fallow, (T) young fallow, (B) new fallow, and (G) extensive rubber gardens. The locations of fruit tree groves and oil plantations are not shown. See Table 2.1 for details about each land-use type

In these plots we investigated the diversity of trees, aphyllophoraceous fungi, small mammals, bats, phytophagous scarabaeid beetles, army ants, and ants that interact mutually with plants and hemipterans. The combination of study plots differed for each investigation (Table 2.2).

Subsequently, biodiversity indices were estimated for each taxonomic group and were compared among those obtained for the different stages of regeneration or land-use types. See the original published articles for detailed information about each taxon, sampling methods, and estimation of biodiversity indices.

Table 2.2 Number of study plots for each land-use type in each of the studies, presenting the diversity indices affected by swidden cultivation

Taxon/biological group	Land-use type								Diversity indices affected by swidden cultivation	References
	New fallow	Young fallow	Old fallow	Fragmented primary forest	Continuous primary forest	Extensive rubber plantation	Fruit tree grove	Oil palm plantation		
Trees		10	13	9	9	5			Species diversity, species composition, and the abundance of rare, endemic, and upper-layer trees	Nakagawa et al. (2013)
Aphyllophoraceous fungi		2	2	2	2	2			Species density and species composition in new fallows	Yamashita et al. (2008)
Small mammals	3	3	3	6	2	3			Species composition in new fallows, but not for the other land-use types	Nakagawa et al. (2006)
Bats	1	1	1		6		3	5	Simpson's index of diversity	Fukuda et al. (2009)
Phytophagous scarabaeids	3	3	3	5	3	3			Evenness and species composition in new fallows, but not for the other land-use types	Kishimoto-Yamada et al. (2011)
Army ants	2	2	2	2	2				Encounter rates	Matsumoto et al. (2009)
Trees with extrafloral nectaries	4	5	6		6				Species richness	Tanaka et al. (2007)
Trees with hemipterans	4	5	6		6				Species richness	Tanaka et al. (2007)
Trees with hemipteran-attending ants	4	5	6		6				Species richness	Tanaka et al. (2007)

(continued)

Table 2.2 (continued)

Taxon/biological group	Land-use type: New fallow	Young fallow	Old fallow	Fragmented primary forest	Continuous primary forest	Extensive rubber plantation	Fruit tree grove	Oil palm plantation	Diversity indices affected by swidden cultivation	References
Macaranga myrmecophyte	4	5	6		6				Species richness, frequency	Tanaka et al. (2007)
Extrafloral nectaries-attending ants	4	5	6		6				Species richness	Tanaka et al. (2007)
Hemipteran-attending ants	4	5	6		6				Species richness	Tanaka et al. (2007)
Tree with weaver ants	4	5	6		6				Frequency	Tanaka et al. (2007)

2.3 Recovery of Biodiversity During Fallow Periods and the Effect of Land-Use Changes on the Biodiversity of Each Taxonomic Group

2.3.1 Trees

Forest structure and tree species composition are fundamentally important habitat characteristics for a diverse range of animals and fungi. Nakagawa et al. (2013) conducted a census in 39–46 study plots and recorded 2,828 trees, representing 552 species from 58 families. Subsequently, the authors analyzed tree community structure (including trunk density, trunk basal area, tree diversity, abundance of rare, endemic, and upper-layer trees, and species composition).

A comparison of species composition implied that succession progresses steadily during fallow periods. Little distinction in species composition was observed among young and old fallow plots, indicating the relatively quick recovery of the tree community during the early stages of succession. However, the values of stem density, basal area, tree diversity, and the abundance of rare, endemic, and upper-layer trees were smaller in old fallows compared to continuous and fragmented primary forests.

The results of this study indicate that the tree communities of secondary forests had not fully recovered, even after fallow periods of 20–60 years.

2.3.2 Aphyllophoraceous Fungi

Yamashita et al. (2008) collected 155 fruiting bodies, representing 67 species of polypores, hydnoid fungi, and stereoid fungi. The authors reported that the species density (Gotelli and Colwell 2001), which was represented by the number of fungus species per 200 m^2, gradually increased from young fallows, old fallows, fragmented primary forests, and extensive rubber gardens to continuous primary forests. This observation indicates how fungal diversity partly recovers during the process of forest succession. However, the level of species density in old fallows was approximately one-third that of the primary forest.

Several environmental variables were also recorded (including canopy openness, water potential, amount and composition of coarse woody debris, litter mass, tree basal area, and tree species composition), of which the amount of coarse woody debris was significantly and positively correlated with fungal species density. Fungal species density was also correlated with principal component analysis scores for tree species composition. However, only the amount of coarse woody debris significantly affected the species composition of fungi.

2.3.3 Small Mammals

Nakagawa et al. (2006) collected 283 individuals of small mammals, representing 22 species belonging to the orders of Scandentia, Rodentia, and Insectivora. Small mammals in new fallows formed a distinctive group in an analysis of species composition, with a high emergence of human-associated rats and arboreal tree shrews. However, the structure of small mammal communities was similar in the other forest types, indicating that the various forest uses, except for new fallows, do not have a major impact on the structure of small mammal communities.

2.3.4 Bats

Fukuda et al. (2009) collected 495 bats, representing 28 species. The authors compared bat diversity for different land-use types, including the primary forest, fallows (new, young, and old fallows combined), fruit tree groves, and oil palm plantations. Simpson's index of diversity was relatively high in the primary forest (7.86) compared to fallows (3.38), fruit tree groves (3.65), and oil palm plantations (1.24). The authors also investigated the feeding habits of megachiropterans (i.e., large bats) by collecting pollen samples from the heads of captured bats. The authors suggested that many species of bats rarely use fruit tree groves and oil palm plantations for feeding. These results indicate that fruit tree groves and oil palm plantations are not suitable for maintaining bat diversity; however, these habitats may provide important food sources for a few species of megachiropterans.

2.3.5 Phytophagous Scarabaeid Beetles

Kishimoto-Yamada et al. (2011) used light traps and collected 3,230 beetles, which represented at least 51 morpho-species of phytophagous scarabaeids (Coleoptera). Estimated species richness did not differ significantly among forest types. The results of this study indicate that fallows at various stages of succession, except for new fallows, tend to maintain phytophagous scarabaeid diversity at a similar level to that found in the primary forest. The authors suggested the roles of complex mosaic landscapes, containing combinations of continuous and fragmented primary forests and fallows, should be re-evaluated in terms of biodiversity conservation.

2.3.6 Army Ants

Matsumoto et al. (2009) assessed the encounter rates of army ant *Aenictus* spp. (Hymenoptera: Formicidae) colonies present in different forest types. This genus represents one of the top predators in litter-arthropod assemblages. Encounter rates were the highest in continuous and fragmented primary forests, intermediate in old and young fallows, and lowest in new fallows. This finding indicates that the abundance of top leaf-litter predators, which are rare and likely to be vulnerable to disturbance, did not fully recover even after fallow periods of 20 years. In terms of forest management strategies to sustain biodiversity, the study suggests it is necessary to preserve primary forests, rather than extending fallow periods, in tropical forest regions where traditional swidden cultivation is performed.

2.3.7 Mutually Interacting Plants, Ants, and Hemipterans

Tanaka et al. (2007) investigated the frequency of occurrence and composition of species involved in ant-associated mutualistic interactions with plants and hemipterans. Species richness was higher in the primary forest compared to fallows for trees bearing extrafloral nectaries, for trees with hemipteran-attending ants, and for ants attending both extrafloral nectaries and hemipterans. More than 80 % of species observed in the primary forest were not recorded in fallows. Myrmecophyte *Macaranga* spp. exhibited higher species richness and the frequency of occurrence in the primary forest compared to fallows. The number of myrmecophyte *Macaranga* species observed in fallows was approximately one-third of the number in the primary forest. In contrast, the weaver ant, *Oecophylla smaragdina*, which tends to exclude other arboreal ant species, was significantly more abundant in fallows compared to the primary forest. These results indicate swidden cultivation drastically decreases the diversity of species involved in ant-plant or ant-hemipteran mutualistic interactions.

2.4 Discussion

2.4.1 Recovery of Biodiversity During the Reforestation Process and the Effects of Swidden Cultivation on Biodiversity

In the current study, species richness, species density, and encounter rates increased from new (or young) to old fallows for trees (Nakagawa et al. 2013), fungi (Yamashita et al. 2008), army ants (Matsumoto et al. 2009), and ants attending hemipterans (Tanaka et al. 2007). This trend was also observed for forest-edge-dwelling butterflies (Itioka et al. 2014) and beetles that use coarse woody debris.

In contrast, even after fallow periods of 20–60 years, the levels of species diversity in old secondary forests had only recovered to approximately one-third to two-third of that recorded in continuous and fragmented primary forests for trees (Nakagawa et al. 2013), fungi (Yamashita et al. 2008), army ants (Matsumoto et al. 2009), trees bearing extrafloral nectaries (Tanaka et al. 2007), and ants attending both extrafloral nectaries and hemipterans (Tanaka et al. 2007). These trends were also observed for forest-edge-dwelling butterflies (Itioka et al. 2014) and beetles that use coarse woody debris (Takano et al., unpublished data). These results indicate that swidden cultivation has significant impact on the biodiversity of the tropical forest region in Borneo. Preserving primary forests is essential to sustain the biodiversity of landscape mosaics created from the crop-fallow cycles of swidden cultivation.

Chazdon et al. (2009) hypothesized that the conservation of primary-forest species in secondary forests would be maximized in older secondary forests that have persisted where anthropogenic disturbance after abandonment is relatively low. The authors also suggested how the conservation value of such secondary forests would increase over time, due to the accumulation of species arriving from remaining primary forest patches.

In the practice of swidden cultivation, the fallow-forest phase serves the function of restoring carbon and nutrient stocks in the biomass, improving soil physical properties, and suppressing weeds (Palm et al. 2005). The length of fallow periods has been reported to vary widely from 3 years to more than 25 years in Malaysian Borneo and Indonesian Kalimantan (Freeman 1955; Padoch 1982; Cramb 1993; Mertz and Christensen 1997; Ichikawa 2010). Recently, the length of fallow periods has mainly declined in Southeast Asia (Schmidt-Vogt et al. 2009), while the length of cropping periods has increased in some parts of Sarawak (Ichikawa 2010). This change is probably due to population pressure (Hansen and Mertz 2006), shortage of farmland for certain households (cf. Coomes et al. 2000), and a need for shorter travel distance to the fields (Nielsen et al. 2006; Ichikawa 2010). Shortened fallow periods and elongated cropping periods are realized by the use of herbicides and fertilizers in some parts of Sarawak (Ichikawa 2010). However, the shortening of fallow periods may have further negative impacts on the biodiversity of these secondary forests.

2.4.2 Effects of Other Land-Use Changes on Biodiversity

Hattori et al. (2012) suggested that older secondary forests, reduced-impact logged forests, and matured tree plantations might be of some importance for the conservation of some fungal species. In the current study, the biodiversity indices of aphyllophoraceous fungi in extensive rubber gardens were lower than those in the continuous primary forest, but were higher compared to those in the fallows (Yamashita et al. 2008). This trend was also observed for phytophagous scarabaeid beetles (Kishimoto-Yamada et al. 2011) and beetles that use coarse woody debris (Takano et al., unpublished data). The amount of coarse woody debris was the important environment factor that explained fungal species density in different study plots (Yamashita et al. 2008). In general, the amount of coarse woody debris in the rubber gardens was intermediate between the amounts in fallows and in primary forests. Hence, this environmental factor might explain the relatively higher biodiversity of aphyllophoraceous fungi and beetles using coarse woody debris in the extensive rubber gardens.

Fukuda et al. (2009) reported that the species richness of bats did not differ between fallows (new, young, and old fallows combined) and the continuous primary forest. In comparison, the lowest bat species richness was recorded in oil palm plantations, followed by fruit tree groves. The authors suggested that secondary forests in fallows potentially maintain bat diversity, whereas fruit tree groves and oil palm plantations provide inadequate habitats for most frugivorous and insectivorous bats.

Oil palm plantations are known to support far fewer wildlife species compared to primary forests and often have fewer species compared to other tree crops and secondary forests (Fitzherbert et al. 2008). There were not enough data available for the current study to compare the biodiversity in fallows, rubber gardens, and oil palm plantations. Further studies are necessary to evaluate the biodiversity conservation value of different types of land use to identify better strategies for balancing human activity and biodiversity conservation.

2.4.3 Effects of Fragmentation and Distance from Continuous Primary Forests on Biodiversity

Encounter rates of army ants decreased in fragmented primary forests compared to continuous primary forest (Matsumoto et al. 2009). The species density of aphyllophoraceous fungi was also slightly lower in fragmented primary forests compared to continuous primary forest (Yamashita et al. 2008). Such trends were also observed for the species richness of forest-edge-dwelling butterflies (Itioka et al. 2014) and beetles that use coarse woody debris (Takano et al., unpublished data). These results indicate how the fragmentation of the primary forest has a negative impact on biodiversity.

Recent analyses of the current study have also revealed how increasing distance from the continuous primary forest (i.e., the LHNP) has had a negative effect on species richness in fallows and fragmented primary forests for forest-edge-dwelling butterflies (Itioka et al. 2014) and beetles using coarse woody debris (Takano et al., unpublished data). These results indicate that the continuous primary forest also serves as a source population (MacArthur and Wilson 1967; Mittelbach 2012) for local wildlife populations inhabiting nearby areas.

In comparison, species richness was higher in fragmented primary forests compared to continuous primary forest for trees and phytophagous scarabaeids. Many biodiversity studies, including the current study, are conducted in small plots that are vulnerable to edge effects from adjacent sites. Such study design may have resulted in the overestimation of the biodiversity indices in fragmented primary forests (Barlow et al. 2007).

2.4.4 Diversity in Landscape Mosaics

Nakagawa et al. (2013) verified that tree diversity is maintained across various spatial scales in and around the LHNP, specifically within plots, between plots of the same forest type, and among plots of different forest types. The authors used additive partitioning of diversity, which is a useful tool for understanding the distribution pattern of diversity in a hierarchically structured landscape (Veech and Crist 2010). The highest diversity was found among forest types, indicating that the complete forested landscape comprises a suitable scale for the conservation of tree diversity in the region.

However, there was a significant increase in the numbers of (1) singleton species (i.e., rare species recorded by a single sample or observation), (2) tree species that are endemic to Borneo, and (3) canopy and emergent trees in primary forests compared to fallows and rubber gardens (Nakagawa et al. 2013). Hattori et al. (2012) also suggested the preservation of primary forests is essential for conserving rare wood-inhabiting fungi in Malaysia. If species that are rare, endemic, and restricted to primary forests are considered, biodiversity conservation should focus on primary forest patches in landscape mosaics.

Kishimoto-Yamada et al. (2013) determined how the proportion of fragmented primary forests in a landscape affects the species diversity and species composition of ants and dung beetles in the *Macaranga*-dominated secondary forests of Sarawak. The proportion of fragmented primary forests within a 100-m radius had a significant positive effect on ant species diversity, with fragments within 100-, 300-, and 500-m radii of primary forests significantly affecting species compositions. These findings indicate that ant species diversity could be enhanced in secondary forests by retaining primary forests occurring within a range of 100 m. It is also important to protect primary and secondary forests at larger spatial scales, such as in range of 100–1,000 m, which would maximize the species diversity of dung beetles in local areas of the landscape (Kishimoto-Yamada et al. 2013).

2.4.5 Variation Among Taxonomic Groups in Response to Disturbance

For small mammals and phytophagous scarabaeid beetles, species richness was similar, or even slightly lower, in the continuous primary forest compared to fragmented primary forests, fallows, and rubber gardens (Nakagawa et al. 2006;

Kishimoto-Yamada et al. 2011). Changes in land use, forest structure, and tree species composition appear to affect animal communities differently depending on the taxonomic groups (Barlow et al. 2007; Chazdon et al. 2009).

Barlow et al. (2007) compared the biodiversity indices of tropical primary, secondary, and plantation forests. The authors reported that different taxa vary markedly in their response to patterns of land use, in terms of species richness and the percentage of species restricted to primary forests (ranging from 5 % to 57 %); however, almost all between-forest comparisons showed marked differences in community structure and composition. Gibson et al. (2011) conducted a meta-analysis of 138 studies on tropical forests. The authors reported that biodiversity indices were substantially lower in degraded forests, but that these indices varied considerably with respect to geographic region, taxonomic group, and disturbance type.

Such variation or discrepancies were also observed for the results of the current study compared to Gibson et al. (2011). For example, Gibson et al. (2011) reported that Coleoptera are more sensitive to disturbance compared to Hymenoptera and Lepidoptera. However, in the current study, army ants (Matsumoto et al. 2009) and forest-edge-dwelling butterflies (Itioka et al. 2014) were sensitive to land cover changes, whereas phytophagous scarabaeid beetles (Kishimoto-Yamada et al. 2011) were less sensitive in terms of encounter rates and species richness.

Butterflies may be roughly classified into herb, liana, and tree specialists and generalists, with different guild probably exhibiting different responses to the same type of disturbance (Cleary et al. 2005). In the case of Coleoptera, phytophagous scarabaeids, which are generalists preferentially feeding on newly emerged leaves, may be attracted to new fallows with young trees, whereas beetles using coarse woody debris are more likely to prefer primary forests and old fallows. These results indicate that we must consider the biological characteristics (e.g., plasticity in feeding habits, dispersal ability, and life history traits) which affect differences in the disturbance tolerance of various species.

2.4.6 *Disturbance Tolerance Through Plasticity in Feeding Habits*

Gibson et al. (2011) reported that mammals are less sensitive to disturbance and, in some instances, actually benefit from human disturbance. The authors suggested some mammals might have a high tolerance to degraded forests and forest edges (Daily et al. 2003), particularly small mammals and bats which dominate most mammal studies.

To evaluate the trophic levels of small mammals, Nakagawa et al. (2007) compared the stable isotope $\delta^{15}N$ values of the toes of small mammals collected in fallows and primary forests. Higher $\delta^{15}N$ values indicate higher trophic levels (Minagawa and Wada 1984; Deniro and Epstein 1981). The $\delta^{15}N$ values for tree shrews and squirrels were similar among land-use types. However, the $\delta^{15}N$ values of rats and mice in new and young fallows were more enriched compared to those in old fallows and primary forests. This increase in ^{15}N indicates that rats and mice inhabiting areas at the early stage of forest succession are likely to feed on more consumers, such as insects, than plant organs compared to rodents inhabiting old fallows or primary forests. These results indicate that some rats and mice are able to adjust their diet and, thus, are potentially tolerant to the conditions in degraded forests.

2.5 Future Studies

The current study, along with previous studies (e.g., Barlow et al. 2007; Chazdon et al. 2009; Gibson et al. 2011), highlights the fundamental issues associated with quantifying biodiversity in anthropogenically altered habitats. Consequently, more data on the characteristics of each taxon (e.g., habitat range, plasticity in feeding habits, dispersal ability, and life history traits) and the types of anthropogenic disturbance (intensity, spatio-temporal configuration, and consequent changes in habitat environment) are required (e.g., Liow et al. 2001; Yamashita et al. 2008). Access to such information would help toward understanding the ecological mechanisms underlying the varying vulnerability of different taxonomic groups to different types of human disturbance. Finally, it is also important to investigate the relationships between the changes in biodiversity and ecosystem functions and dynamics (e.g., decomposition services and nutrient cycles, natural enemy services and agricultural productivity, pollination services and plant reproductive success, and dispersal services and forest succession).

References

Ashton PS (2005) Lambir's forest: the world's most diverse known tree assemblage? In: Roubik DW, Sakai S, Hamid Karim AA (eds) Pollination ecololgy and the rain forest: Sarawak studies. Springer, New York, pp 191–216

Barlow J, Gardner TA, Araujo IS, Avila-Pires TC, Bonaldo AB, Costa JE, Esposito MC, Ferreira LV, Hawes J, Hernandez MM, Hoogmoed MS, Leite RN, Lo-Man-Hung NF, Malcolm JR, Martins MB, Mestre LAM, Miranda-Santos R, Nunes-Gutjahr AL, Overal WL, Parry L, Peters SL, Ribeiro-Junior MA, da Silva MNF, Motta CD, Peres CA (2007) Quantifying the biodiversity value of tropical primary, secondary, and plantation forests. Proc Natl Acad Sci U S A 104(47):18555–18560. doi:10.1073/pnas.0703333104

Chazdon RL, Peres CA, Dent D, Sheil D, Lugo AE, Lamb D, Stork NE, Miller SE (2009) The potential for species conservation in tropical secondary forests. Conserv Biol 23(6):1406–1417. doi:10.1111/j.1523-1739.2009.01338.x

Cleary DFR, Boyle TJB, Setyawati T, Menken SBJ (2005) The impact of logging on the abundance, species richness and community composition of butterfly guilds in Borneo. J Appl Entomol 129(1):52–59. doi:10.1111/j.1439-0418.2005.00916.x

Condit R, Ashton PS, Baker P, Bunyavejchewin S, Gunatilleke S, Gunatilleke N, Hubbell SP, Foster RB, Itoh A, LaFrankie JV, Lee HS, Losos E, Manokaran N, Sukumar R, Yamakura T (2000) Spatial patterns in the distribution of tropical tree species. Science 288(5470):1414–1418. doi:10.1126/science.288.5470.1414

Coomes OT, Grimard F, Burt GJ (2000) Tropical forests and shifting cultivation: secondary forest fallow dynamics among traditional farmers of the Peruvian Amazon. Ecol Econ 32(1):109–124. doi:10.1016/s0921-8009(99)00066-x

Corlett R (2009) The ecology of tropical East Asia. Oxford biology. Oxford University Press, New York

Cramb RA (1993) Shifting cultivation and sustainable agriculture in East Malaysia: a longitudinal case-study. Agric Syst 42(3):209–226. doi:10.1016/0308-521x(93)90055-7

Daily GC, Ceballos G, Pacheco J, Suzan G, Sanchez-Azofeifa A (2003) Countryside biogeography of neotropical mammals: conservation opportunities in agricultural landscapes of Costa Rica. Conserv Biol 17(6):1814–1826. doi:10.1111/j.1523-1739.2003.00298.x

Debinski DM, Holt RD (2000) A survey and overview of habitat fragmentation experiments. Conserv Biol 14(2):342–355. doi:10.1046/j.1523-1739.2000.98081.x

Deniro MJ, Epstein S (1981) Influence of diet on the distribution of nitrogen isotopes in animals. Geochim Cosmochim Acta 45(3):341–351. doi:10.1016/0016-7037(81)90244-1

Department of Statistics Malaysia (2004) Yearbook of statistics Malaysia 2004. Putrajaya, Malaysia

Dove MR (1985) Swidden agriculture in Indonesia: the subsistence strategies of the Kalimantan Kantu. Mouton Publishers, Berlin

FAO (2010) Global forest resources assessment 2010. Food and Agriculture Organization of the United Nations, Rome

Fitzherbert EB, Struebig MJ, Morel A, Danielsen F, Bruhl CA, Donald PF, Phalan B (2008) How will oil palm expansion affect biodiversity? Trends Ecol Evol 23(10):538–545. doi:10.1016/j.tree.2008.06.012

Freeman JD (1955) Iban agriculture: a report on the shifting cultivationof hill rice by the Iban of Sarawak. H.M.S.O, London

Fukuda D, Tisen OB, Momose K, Sakai S (2009) Bat diversity in the vegetation mosaic around a lowland dipterocarp forest of Borneo. Raffles Bull Zool 57(1):213–221

Gibson L, Lee TM, Koh LP, Brook BW, Gardner TA, Barlow J, Peres CA, Bradshaw CJ, Laurance WF, Lovejoy TE, Sodhi NS (2011) Primary forests are irreplaceable for sustaining tropical biodiversity. Nature 478(7369):378–381. doi:10.1038/nature10425

Gotelli NJ, Colwell RK (2001) Quantifying biodiversity: procedures and pitfalls in the measurement and comparison of species richness. Ecol Lett 4(4):379–391. doi:10.1046/j.1461-0248.2001.00230.x

Hansen TS, Mertz O (2006) Extinction or adaptation? Three decades of change in shifting cultivation in Sarawak, Malaysia. Land Degrad Dev 17(2):135–148. doi:10.1002/ldr.720

Hattori T, Yamashita S, Lee SS (2012) Diversity and conservation of wood-inhabiting polypores and other aphyllophoraceous fungi in Malaysia. Biodivers Conserv 21(9):2375–2396. doi:10.1007/s10531-012-0238-x

Hazebroek HP, Kashim bin Abang Morshidi A (2000) National parks of Sarawak. National History Publications (Borneo), Kota Kinabalu, Malaysia

Ichikawa M (2003) One hundred years of land-use changes: political, social, and economic influences on an Iban village in Bakong river basin, Sarawak, East Malaysia. In: Tuck-Po L, de Jong W, Abe K (eds) The political ecology of tropical forests in southeast Asia: historical perspectives. Kyoto University Press/Trans Pacific Press, Kyoto, pp 177–199

Ichikawa M (2007) Degradation and loss of forest land and land-use changes in Sarawak, East Malaysia: a study of native land use by the Iban. Ecol Res 22(3):403–413. doi:10.1007/s11284-007-0365-0

Ichikawa M (2010) Malaysia, Sarawak shuu no yakihata saibai ni mirareru josouzai riyou to sono haikei (utilization of herbicide and its background in swidden cultivation in the state of Sarawak, Malaysia). Noko no Gijutsu to Bunka 27:21–41 (in Japanese)

Itioka T, Yamamoto T, Tzuchiya T, Okubo T, Yago M, Seki Y, Ohshima Y, Katsuyama R, Chiba H, Yata O (2009) Butterflies collected in and around Lambir Hills National Park, Sarawak, Malaysia in Borneo. Contributions from the Biological Laboratory, Kyoto University 30(1):25–68

Itioka T, Takano KT, Kishimo-Yamada K, Tzuchiya T, Ohshima Y, Katsuyama R-i, Yago M, Nakagawa M, Nakashizuka T (2014) Chronosequential changes in species richness of forest-edge-dwelling butterflies during forest restoration after swidden cultivation in a humid tropical rainforest region in Borneo. J For Res. doi:10.1007/s10310-014-0444-3

Kato M, Inoue T, Hamid AA, Nagamitsu T, Merdek MB, Nona AR, Itino T, Yamane S, Yumoto T (1995) Seasonality and vertical structure of light-attracted insect communities in a dipterocarp forest in Sarawak. Res Popul Ecol 37(1):59–79. doi:10.1007/bf02515762

Kishimoto-Yamada K, Itioka T, Nakagawa M, Momose K, Nakashizuka T (2011) Phytophagous scarabaeid diversity in Swidden cultivation landscapes in Sarawak, Malaysia. Raffles Bull Zool 59(2):285–293

Kishimoto-Yamada K, Hyodo F, Matsuoka M, Hashimoto Y, Kon M, Ochi T, Yamane S, Ishii R, Itioka T (2013) Effects of remnant primary forests on ant and dung beetle species diversity in a secondary forest in Sarawak, Malaysia. J Insect Conserv. doi:10.1007/s10841-012-9544-6

Kume T, Tanaka N, Kuraji K, Komatsu H, Yoshifuji N, Saitoh TM, Suzuki M, Kumagai T (2011) Ten-year evapotranspiration estimates in a Bornean tropical rainforest. Agr Forest Meteorol 151(9):1183–1192. doi:10.1016/j.agrformet.2011.04.005

Lanly JP (1982) Tropical forest resources. FAO Forestry Paper 30. Food and Agriculture Organization of the United Nations, Rome

Liow LH, Sodhi NS, Elmqvist T (2001) Bee diversity along a disturbance gradient in tropical lowland forests of south-east Asia. J Appl Ecol 38(1):180–192. doi:10.1046/j.1365-2664.2001.00582.x

MacArthur RH, Wilson EO (1967) The theory of island biogeography. Princeton University Press, Princeton

Matsumoto T, Itioka T, Yamane S, Momose K (2009) Traditional land use associated with swidden agriculture changes encounter rates of the top predator, the army ant, in Southeast Asian tropical rain forests. Biodivers Conserv 18(12):3139–3151. doi:10.1007/s10531-009-9632-4

Mertz O, Christensen H (1997) Land use and crop diversity in two Iban communities, Sarawak, Malaysia. Geografisk Tidsskrift, Danish J Geogr 97:98–110

Mertz O, Leisz SJ, Heinimann A, Rerkasem K, Thiha DW, Pham VC, Vu KC, Schmidt-Vogt D, Colfer CJP, Epprecht M, Padoch C, Potter L (2009a) Who counts? Demography of swidden cultivators in Southeast Asia. Hum Ecol 37(3):281–289. doi:10.1007/s10745-009-9249-y

Mertz O, Padoch C, Fox J, Cramb RA, Leisz SJ, Lam NT, Vien TD (2009b) Swidden change in Southeast Asia: understanding causes and consequences. Hum Ecol 37(3):259–264. doi:10.1007/s10745-009-9245-2

Minagawa M, Wada E (1984) Stepwise enrichment of ^{15}N along food chains: further evidence and the relation between δ^{15}N and animal age. Geochim Cosmochim Acta 48(5):1135–1140. doi:10.1016/0016-7037(84)90204-7

Mittelbach GG (2012) Community ecology. Sinauer Associates, Sunderland

Nakagawa M, Miguchi H, Nakashizuka T (2006) The effects of various forest uses on small mammal communities in Sarawak, Malaysia. For Ecol Manage 231(1–3):55–62. doi:10.1016/j.foreco.2006.05.006

Nakagawa M, Hyodo F, Nakashizuka T (2007) Effect of forest use on trophic levels of small mammals: an analysis using stable isotopes. Can J Zool Rev Can Zool 85(4):472–478. doi:10.1139/z07-026

Nakagawa M, Momose K, Kishimoto-Yamada K, Kamoi T, Tanaka HO, Kaga M, Yamashita S, Itioka T, Nagamasu H, Sakai S, Nakashizuka T (2013) Tree community structure, dynamics, and diversity partitioning in a Bornean tropical forested landscape. Biodivers Conserv 22(1):127–140. doi:10.1007/s10531-012-0405-0

Ngidang D (2005) Deconstruction and reconstruction of native customary land tenure in Sarawak. Southeast Asia Stud 43(1):47–75
Nielsen U, Mertz O, Noweg GT (2006) The rationality of shifting cultivation systems: labor productivity revisited. Hum Ecol 34(2):201–218. doi:10.1007/s10745-006-9014-4
Padoch C (1982) Migration and its alternatives among the Iban of Sarawak. Marrtinus Nijihoff, The Hague/Leiden
Padoch C, Pinedo-Vasquez M (2010) Saving slash-and-burn to save biodiversity. Biotropica 42 (5):550–552. doi:10.1111/j.1744-7429.2010.00681.x
Palm CA, Vosti SA, Sanchez PA, Ericksen PJ (eds) (2005) Slash-and-burn agriculture: the search for alternatives. Columbia University Press, New York
Pringle R (1970) Rajahs and rebels: the Iban of Sarawak under Brooke rule, 1841-1941. Cornell University Press, Ithaca
Rerkasem K, Lawrence D, Padoch C, Schmidt-Vogt D, Ziegler AD, Bruun TB (2009) Consequences of swidden transitions for crop and fallow biodiversity in Southeast Asia. Hum Ecol 37 (3):347–360. doi:10.1007/s10745-009-9250-5
Samejima H, Ong R, Lagan P, Kitayama K (2012) Camera-trapping rates of mammals and birds in a Bornean tropical rainforest under sustainable forest management. For Ecol Manage 270:248–256. doi:10.1016/j.foreco.2012.01.013
Sandin B (1967) The Sea Dayak of Borneo before white Rajah rule. Michigan State University Press, East Lausing
Sandin B (1994) Sources of Iban traditional history. The Sarawak Museum Journal 67 (special monograph no 7)
Schmidt-Vogt D, Leisz SJ, Mertz O, Heinimann A, Thiha T, Messerli P, Epprecht M, Cu PV, Chi VK, Hardiono M, Dao TM (2009) An assessment of trends in the extent of swidden in Southeast Asia. Hum Ecol 37(3):269–280. doi:10.1007/s10745-009-9239-0
Tanaka HO, Yamane S, Nakashizuka T, Momose K, Itioka T (2007) Effects of deforestation on mutualistic interactions of ants with plants and hemipterans in tropical rainforest of Borneo. Asia Myrmecol 1:31–50
Tscharntke T, Leuschner C, Zeller M, Guhardja E, Bidin A (eds) (2007) Stability of tropical rainforest margins: linking ecological, economic and social constraints of land use and conservation. Springer, New York
Veddeler D, Schulze CH, Steffan-Dewenter I, Buchori D, Tscharntke T (2005) The contribution of tropical secondary forest fragments to the conservation of fruit-feeding butterflies: effects of isolation and age. Biodivers Conserv 14(14):3577–3592. doi:10.1007/s10531-004-0829-2
Veech JA, Crist TO (2010) Toward a unified view of diversity partitioning. Ecology 91(7):1988–1992. doi:10.1890/09-1140.1
Wadley RL, Colfer CJP, Hood IG (1997) Hunting primates and managing forests: the case of Iban forest farmers in Indonesian Borneo. Hum Ecol 25(2):243–271. doi:10.1023/a:1021926206649
Watson H (1985) Lambir Hills National Park: resource inventory with management recommendations. National Parks and Wildlife Office, Forest Department, Kuching, Sarawak, Malaysia
Whitmore TC (1998) An introduction to tropical rain forests, 2nd edn. Oxford University Press, New York
Yamashita S, Hattori T, Momose K, Nakagawa M, Aiba M, Nakashizuka T (2008) Effects of forest use on aphyllophoraceous fungal community structure in Sarawak, Malaysia. Biotropica 40 (3):354–362. doi:10.1111/j.1744-7429.2007.00366.x

Chapter 3
Mongolian Nomadism and the Relationship Between Livestock Grazing, Pasture Vegetation, and Soil Alkalization

Ryosuke Koda and Noboru Fujita

Abstract Grassland covers approximately 75 % of Mongolia, which is characterized by a dry, cold climate, and most of this grassland is used as pasture. Although they are vulnerable, people have lived in the grasslands as nomadic herders for thousands of years, suggesting that the relationship between livestock grazing and pasture vegetation has historically been sustainable. Since the twentieth century, however, the social system and the nomadic circumstances of herders have changed dramatically through a rapid shift to socialism and democracy. In recent decades, the number of livestock (particularly goats) raised to produce cashmere for export has increased rapidly, and herders have concentrated in areas surrounding the capital city and along main roads leading to the capital. These changes, which reflect a shift from nomadic traditions to unsustainable practices, have resulted in the degradation of pasture vegetation. Large herbivores, including livestock, have significant direct and indirect effects on ecosystem structure and function. However, livestock products are an essential part of the Mongolian economy and the livelihood of large numbers of people. It is therefore important to assess the direct and indirect effects of livestock grazing on pasture ecosystems to promote both conservation of natural ecosystems and sustainability of residents and their livelihoods. To obtain a better understanding of the direct effects of livestock grazing, I have used the fecal accumulation rate technique to estimate livestock density on a small scale, and to estimate livestock grazing rates. The data obtained in this study suggest that grazing rate is generally constant, but that it can vary depending on total production of

R. Koda (✉)
Research Institute for Humanity and Nature, 457-4 Motoyama, Kamigamo, Kita-ku, Kyoto 603-8047, Japan

Research Institute of Environment, Agriculture and Fisheries, Osaka Prefecture, 442 Shakudo, Habikino, Osaka 583-0862, Japan
e-mail: KoudaR@mbox.kannousuiken-osaka.or.jp; morinokoda@gmail.com

N. Fujita
Research Institute for Humanity and Nature, Kyoto, Japan

S. Sakai and C. Umetsu (eds.), *Social–Ecological Systems in Transition*, Global Environmental Studies, DOI 10.1007/978-4-431-54910-9_3, © Springer Japan 2014

pasture plants. To understand better the indirect effects of grazing, relationships between soil alkalization, livestock grazing pressure, and pasture vegetation were examined. I conclude that overgrazing can cause soil alkalization, and that soil alkalization may delay the recovery of degraded pasture vegetation. Thus, it is important to consider the variability of the grazing rate and to manage grazing intensity and prevent soil alkalization to achieve sustainable use of pasture.

Keywords Fecal pellet count • Grazing rate • Large herbivore • Nomadic herding • Overgrazing

3.1 Introduction

3.1.1 Environment and Nomadic Herding in Mongolia

Mongolia is a land-locked country located in northeastern Asia, occupying 1.56 million km^2 with a population of approximately 2.8 million (National Statistical Office of Mongolia 2012). Most parts of the country are arid and semi-arid highlands; annual precipitation exceeds 400 mm in northern areas, and is less than 100 mm in southern areas (Fig. 3.1). Precipitation is concentrated during the summer months; less than 10 % of the annual total precipitation occurs during the cold season. Mean annual temperature is higher in southern than in northern Mongolia (Fig. 3.2). January is the coldest month, with the mean temperature ranging from −30 °C in the north to −15 °C in the south (Morinaga et al. 2003). In accord with these latitudinal trends in precipitation and temperature, vegetation changes from north to south. Plant communities include northern coniferous forest, forest-steppe, steppe, desert steppe, and southern desert (Wallis de Vris et al. 1996; Fig. 3.3), forming continuous ecotones (transitional areas between adjacent ecological communities) (Peters 2002). Ecotones are generally sensitive to external

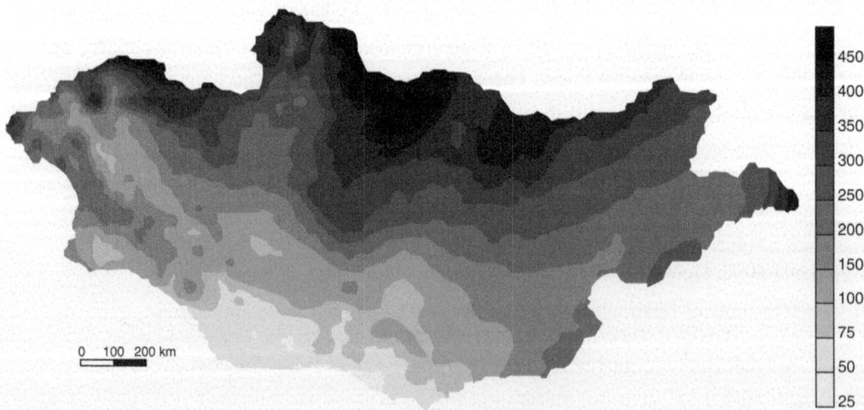

Fig. 3.1 Annual rainfall distribution (mm) in Mongolia. Data from MNET (2009)

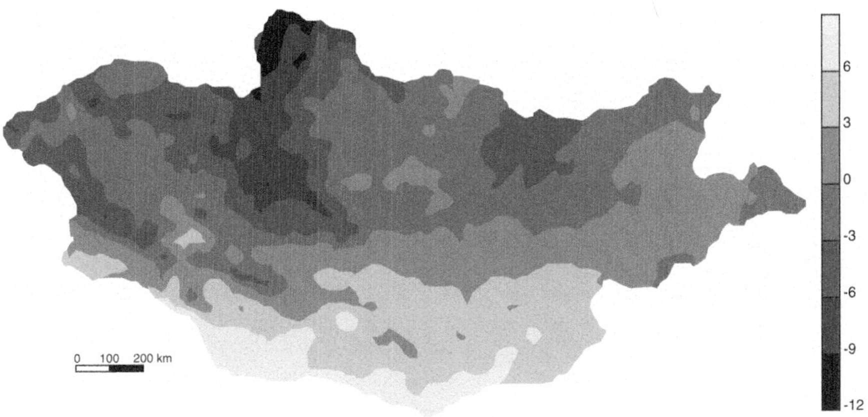

Fig. 3.2 Mean annual temperature (°C) in Mongolia. Data from MNET (2009)

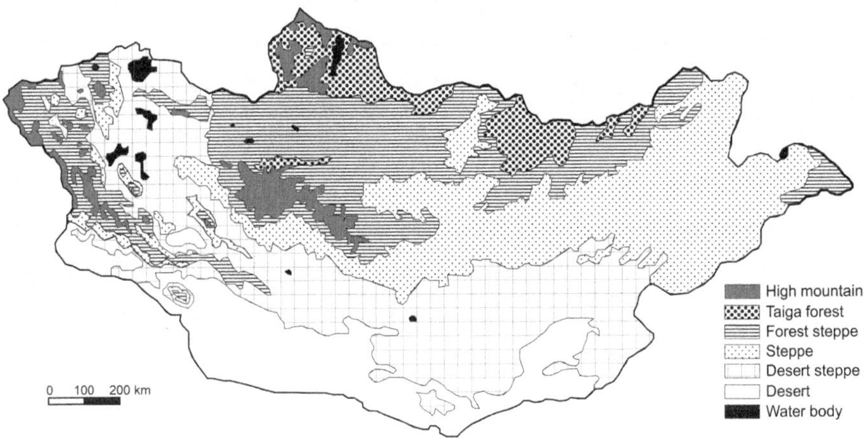

Fig. 3.3 Map of Mongolian biomes (modified from Finch 1999)

environmental disturbances, both natural and anthropogenic, such as climate change, livestock management, and large scale logging (Peters 2002; Pogue and Schnell 2001). Approximately 75 % of the total area of Mongolia is grassland, and most of this area is used as pasture (Fernandez-Gimenez and Allen-Diaz 1999). Most Mongolian peoples have been nomadic, herding cattle, camels, horses, sheep, and goats (Fernandez-Gimenez and Allen-Diaz 1999; Fernandez-Gimenez 2002). Traditionally, nomadic peoples changed their camps (called "gers"; Fig. 3.4) seasonally, depending on topography, water sources, and pasture conditions (Fernandez-Gimenez and Allen-Diaz 1999; Fernandez-Gimenez 2002; Okayasu et al. 2010). Although vulnerable vegetation ecotones occur in Mongolia, nomadic herding persisted for thousands of years based on the traditional ecological knowledge of the herders (Fernandez-Gimenez 2002), suggesting that there has been a sustainable relationship between livestock grazing and pasture vegetation.

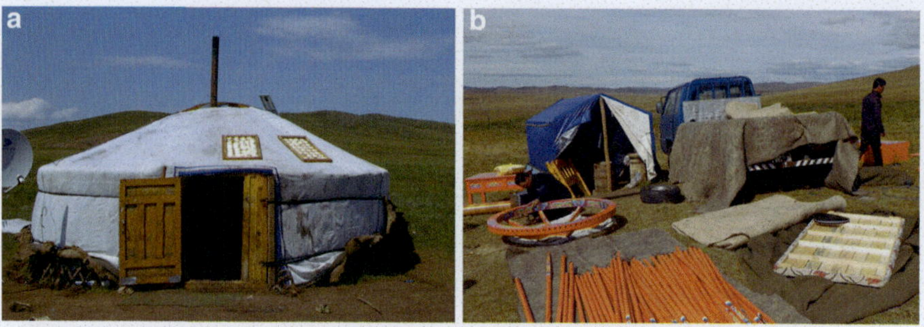

Fig. 3.4 Nomadic camp (**a**) standing in the grassland, and (**b**) assembled after seasonal migration

3.1.2 Changes in Mongolian Social Systems After the Twentieth Century

Since the start of the twentieth century, the social system and nomadic circumstances of herders such as livestock number has changed dramatically. From 1924 to 1990, Mongolia operated under a Soviet-influenced socialist government with a centrally planned socialist economy (Fernandez-Gimenez 2002). By 1960, all herders had joined livestock collectives (called "Negdels"). Most livestock became state owned and the number of livestock was controlled to some extent (Honeychurch 2010). Statistical data show that the number of each species of livestock was almost constant during the socialist era (Fig. 3.5). During this period, the population of Mongolia gradually increased in Ulaanbaatar (the capital city) and surrounding areas (extra-Ulaanbaatar) (Fig. 3.6). To increase agricultural production in response to the increasing population, the government initiated a program of agricultural development from 1959 to 1965 and from 1976 to 1988 (called "atar") (Konagaya 2010), and the total area of agricultural land increased rapidly until 1990 (Fig. 3.7).

Following Mongolia's first democratic elections in 1990, the country transitioned to a free-market economy, and the Negdels were dismantled (Fernandez-Gimenez 2002). All livestock owned by Negdels were transferred to herders as a result of privatization, and decisions about livestock management (e.g., number and composition of stock) were made by individual herders rather than by the state. In addition, many people who worked in Negdels became unemployed shortly after the transition to a democratic system, and most of these individuals became nomadic herders. The statistical data show a rapid increase in the number of herder households in the early 1990s (Fig. 3.7), which was accompanied by a dramatic increase in the number of goats and other livestock (Rossabi 2005), despite mass mortality from cold damage between 1999 and 2002 (Lise et al. 2006; Fig. 3.5). On the other hand, agricultural development programs implemented by the socialist government were suspended, and most of the agricultural land was abandoned (Fig. 3.7). Under the market economy, freed from restrictions on immigration, herders attempted to secure high profitability and concentrated their land use in areas surrounding the capital city and along main

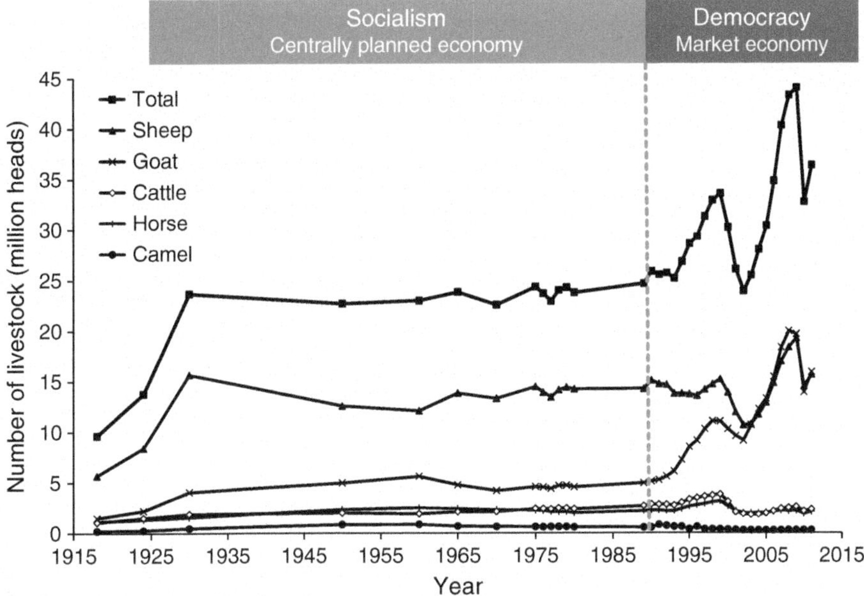

Fig. 3.5 Changes in numbers of livestock by species from 1918 to 2011 in Mongolia. Data from Central Statistical Board under the Council of Ministers of the MPR (1981) and National Statistical Office of Mongolia (2004, 2012)

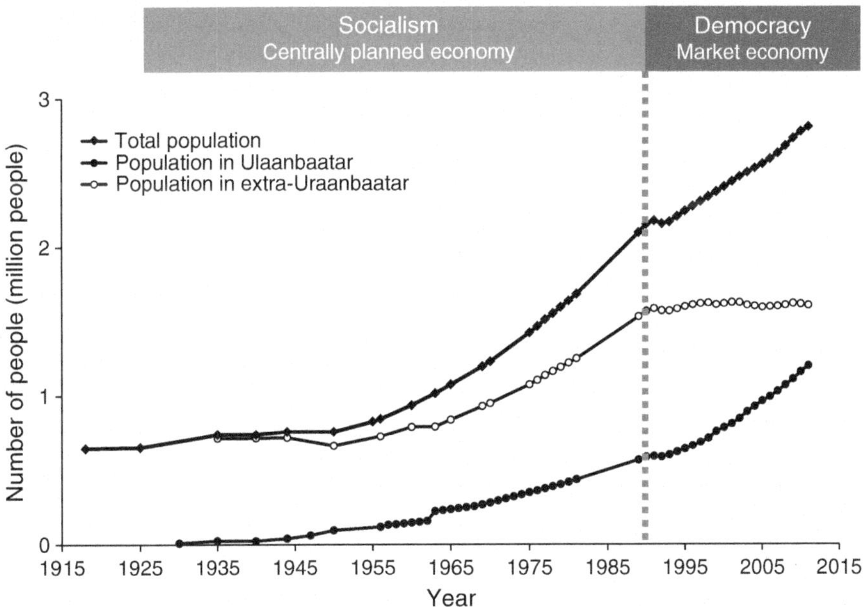

Fig. 3.6 Changes in human population from 1918 to 2011 in the capital and surrounding areas in Mongolia. Data sources as in Fig. 3.5

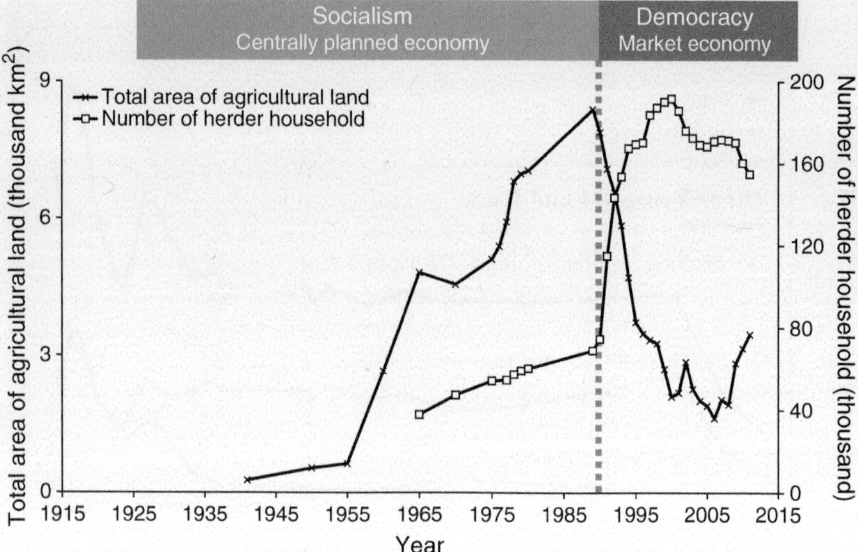

Fig. 3.7 Changes in agricultural land area and number of herder households in Mongolia from 1941 to 2011 and from 1965 to 2011, respectively. Data sources as in Fig. 3.5

roads to the capital, where the price of livestock products was exceptionally high (Okayasu et al. 2007; Honeychurch 2010; Saizen et al. 2010). While the population in extra-Ulaanbaatar has remained constant since about 1990, the population in Ulaanbaatar has continued to increase during the democratic era (Fig. 3.6).

Several problems have arisen in Mongolian ecosystems as a result of these changes in nomadic circumstances, including degradation of pasture and forest vegetation, desertification, and decrease of wild ungulates (Fernandez-Gimenez 2002; Tsogtbaatar 2004; Okayasu et al. 2007, 2010; Yoshihara et al. 2008; Fujita et al. 2012b). The traditional sustainable relationship between livestock grazing and pasture vegetation has been lost due to the mismatch between pasture production and livestock grazing pressure.

3.1.3 Effects of Livestock on Grassland Ecosystems

Large herbivores, including livestock, can have significant effects on ecosystems through their grazing, browsing, trampling, defecation, and urination (Gill 1992; Hobbs 1996; Rooney and Waller 2003; Côté et al. 2004; Mysterud 2006). For example, grazing and browsing by large herbivores has been shown to alter plant species composition and productivity in several ecosystems (Gill and Beardall 2001; Côté et al. 2004; Tremblay et al. 2007; Fujita et al. 2009). Large herbivores are also major determinants of nutrient cycling, directly through feeding and deposition of excrements, and indirectly by modifying the quality and quantity of plant litter

(Hobbs 1996; Ritchie et al. 1998). Large herbivores can affect the distribution of other animal species indirectly, via modification of the composition and physical structure of the habitat (Rooney and Waller 2003; Côté et al. 2004). Thus, large herbivores have pronounced effects on ecosystem structure and function, both directly and indirectly (Rooney and Waller 2003; Côté et al. 2004; Melis et al. 2006).

In Mongolia, livestock play a more important role in pasture ecosystems than do wild ungulates. There are more than 30 million individual livestock in Mongolia, while wild ungulates such as red deer and wild ass have decreased in number during the last century (Reading et al. 2001; Wingard and Zahler 2006). In addition, livestock products are essential to the Mongolian economy and livelihoods of many people (Fernandez-Gimenez 2002). Although the proportion is declining from year to year, livestock products still constitute approximately 15 % of gross domestic product and form one of the most important industries in Mongolia (National Statistical Office of Mongolia 2012). Therefore, it is important to assess the direct and indirect effects of livestock grazing on pasture ecosystems to facilitate conservation of natural ecosystems and sustainability of local residents.

In this chapter, I suggest two important issues that are central to improving our understanding of the effects of livestock on pasture ecosystems. First, it is difficult to ascertain livestock density on small scales from available statistical data, and an established method for estimating small-scale livestock density which could be compared with vegetation survey in quadrats and line transects is currently lacking. This information is critical to a better understanding of the direct impacts of these animals on vegetation. I used the fecal accumulation rate technique (Koda et al. 2011), a fecal pellet count method, to estimate livestock density on a small scale, and estimated livestock grazing rate (the mass of vegetation eaten by one animal in a day) from this density. The second issue concerns the indirect effects of livestock grazing, particularly on soil environments. In Mongolian pastureland, vegetation that has been degraded by overgrazing recovers slowly, if at all, despite decreased density of livestock (Fujita et al. 2012b). I hypothesized that overgrazing caused soil alkalization and that this environmental change may delay the recovery of degraded pastureland.

3.2 Method of Estimating Small-Scale Livestock Density and Grazing Rate

3.2.1 Background

Accurate estimation of population density of large herbivores is essential to achieving a better understanding of the effects of these animals on plant species, and for devising effective management strategies for both herbivores and vegetation (Marques et al. 2001). In Mongolia, the large-scale (e.g., province- or district-level) population density of livestock can easily be estimated from statistical data. However, it is not possible to estimate accurately small-scale livestock density from these data. Investigations of pasture plants, such as mowing experiments (e.g., Fujita et al. 2009; N. Fujita and E. Ariunbold, Chap. 4), are usually conducted on relatively small scales such as quadrats and line transects; it is important to estimate the population density of livestock on similar scales to understand the relationship between pasture vegetation and grazing. For example, although the livestock grazing rate is uncertain, it can be calculated by comparing estimated livestock density with total amount of feeding, using mowing experiments. The density of deer can be estimated on small scales using the fecal pellet-group count method (Koda and Fujita 2011), and this method has been widely used in studies of habitat selection by large herbivores (e.g., Hemami et al. 2004). I tested the use of the fecal pellet-group count method to estimate the density of livestock in Mongolia, and to calculate livestock grazing rate. I used the fecal accumulation rate technique (Campbell et al. 2004; Koda et al. 2011) to estimate livestock density on a small scale. Estimated densities were compared with GPS data on livestock movement and with the total amount of feeding, which was calculated using a mowing experiment.

3.2.2 Application of the Fecal Accumulation Rate Technique and Comparison with GPS Data for Domestic Sheep

In July 2010 I established seven transects (50 × 4 m) in Bayan-Unjuul (steppe area). In this site, a herd of sheep have been fitted with GPS logger and has been observed since 2008. I established transects located 100, 250, 400, 550, 700, 850, and 1,000 m from a nomadic camp which included the GPS fitted sheep. The density of GPS points was calculated in each transect area (0–150, 150–300, 300–450, 450–600, 600–750, 750–900, and 900–1,050 m for the respective transects).

On July 7th in 2010 all fecal pellet groups from sheep and goats were removed from each transect. On August 3rd I visited transects again and counted the number of sheep and goat fecal pellet groups that had accumulated in each transect over 27 days. According to the fecal accumulation rate technique, the density of a given animal can be calculated using the number of accumulated fecal pellet groups, the time interval, and the defecation rate of the animal (Koda et al. 2011). Defecation rates of sheep and goats were estimated as 21.4 and 31.8 (the number of pellet groups per animal per day), respectively (Koda et al., unpublished data). However, it is difficult to distinguish between the fecal pellets of sheep and goats. Therefore, I used the average defecation rate of sheep and goats to estimate the total density of each. The relationship between the estimated density of sheep and goats in each transect and the density of GPS points in each area was tested using linear regression analysis.

A positive correlation was found between estimated livestock densities and GPS data (Fig. 3.8). In the linear regression analysis, the estimated density of sheep and goats showed a statistically significant correlation with the density of GPS points ($p = 0.017$, $R^2 = 0.71$). This result suggests that livestock density as estimated

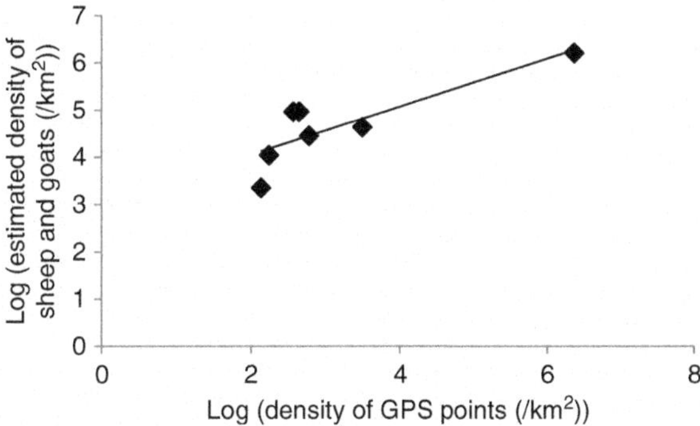

Fig. 3.8 Relationship between the density of GPS points and the estimated density of sheep and goats

using the fecal accumulation rate technique is accurate, and that the fecal accumulation rate technique can thus be applied to estimate livestock density at small scales in Mongolia.

3.2.3 Estimation of Livestock Grazing Rate

To estimate livestock grazing rate, I used estimates of livestock density obtained with the fecal accumulation rate technique, and a mowing experiment that estimated the total amount of feeding. Fenced enclosures (approximately 1 ha) were established in 2008 both in Bayan-Unjuul (steppe) and Erdene (forest steppe), to prevent livestock from grazing on pasture plants. In May 2010, four transects (100×4 m) were established around each fence, and all livestock fecal pellet groups were removed from the transects. In June 2010, fecal pellet groups accumulated in each transect were counted according to animal species (horse, cattle, sheep, and goat) and removed from the transects. This method was repeated in June–July and July–August 2010. I estimated livestock density for each species during each period using the fecal accumulation rate technique. To overcome the difficulty in distinguishing between the fecal pellets of sheep and goats, I estimated total density of sheep and goats using the average defecation rate of each, and divided total density into density of sheep and goats using statistical data on the sheep:goat ratio in Bayan-Unjuul and Erdene.

The total amount of feeding was calculated using a mowing experiment inside the fences. Ten 1×1 m square quadrats were established inside the fence in each area. The height of grazed pasture plants outside each fence was measured randomly at >10 points, and the average grazing height was calculated. To estimate the total amount of feeding, all pasture plants were mowed at the average grazing height in five quadrats inside each fence. To estimate the total production of pasture plants, all pasture plants in the other five quadrats within each fence were mowed at 3 cm, the approximate minimum grazing height (Fujita et al. 2012a). The mowing experiment was conducted during a similar period as the fecal pellet counts (May–June, June–July, and July–August 2010). After each mowing, the trimmed

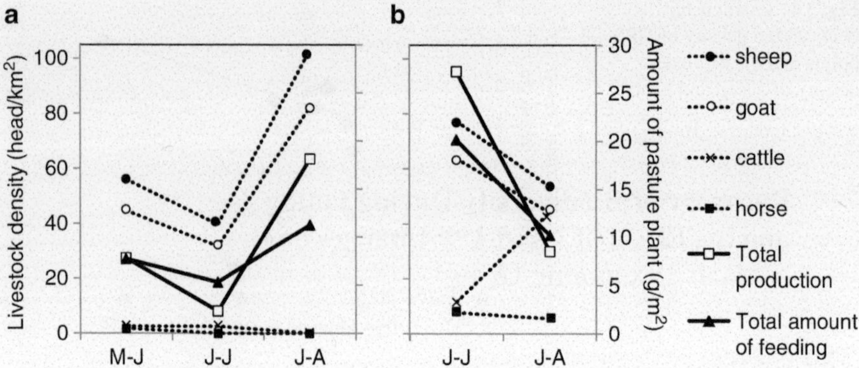

Fig. 3.9 Differences in the estimated density of each livestock species (head/km^2), total production of pasture vegetation (g/m^2), and total amount of feeding (g/m^2) among sampling locations and periods. (**a**) Bayan-Unjuul. (**b**) Erdene. M–J, J–J, and J–A indicate May–June, June–July, and July–August, respectively

pasture plants were transported to the laboratory and dried at 80 °C for 2 days to determine dry mass.

To calculate grazing rate, I converted the livestock density of each species during each sampling period to the sheep unit density, according to the National Statistical Office of Mongolia (see Hoshino et al. 2009): 1 horse = 7 sheep, 1 cattle = 6 sheep, and 1 goat = 0.9 sheep. Grazing rate was calculated as the total amount of feeding per sheep unit density per time interval during the mowing period.

Estimated densities differed among livestock species and study sites. In Bayan-Unjuul, densities of sheep and goats were very high (sheep: 39.6–103.2, goats: 31.6–82.3 head/km^2), while densities of cattle and horses were very low (cattle: 0.00–2.56, horses: 0.00–1.86 head/km^2) (Fig. 3.9a). In Erdene, densities of sheep and goats were high (sheep: 48.8–77.3, goats: 40.2–63.7 head/km^2), and densities of cattle and horses were higher than those in Bayan-Unjuul (cattle: 9.66–44.4, horses: 5.59–11.4 head/km^2) (Fig. 3.9b). The total amount of feeding and pasture production varied between sampling periods. During June–July in Bayan-Unjuul and during July–August in Erdene, total production of pasture plants was lower than the total amount of feeding, which suggests that livestock browsed pasture plants to under 3 cm in height. The variation in total amount of feeding and pasture production was similar to the variation in estimated density of sheep and goats. The grazing rate in Bayan-Unjuul was relatively constant (1.99–2.47 kg/head/day), while grazing rate in Erdene varied between sampling periods (0.91–3.01 kg/head/day).

The grazing rate of livestock is uncertain; it has often been thought that grazing rate is constant, unvarying among locations or months within a season. I found grazing rates were mostly constant, but they differed between steppe and forest steppe areas, and grazing rate in July–August in Erdene was significantly different from other periods. Thus, the grazing rate of livestock can change according to location and time period within a given season. In Erdene, the total production of pasture plants was relatively high, and the grazing rate was higher than in Bayan-Unjuul except for the July–August period, suggesting grazing rate is dependent on the mass of pasture plants produced. The total quantity of vegetation grazed varied according to sheep and goat densities, and did not coincide with cattle density, especially during July–August in Erdene. These results suggest cattle may not be able to consume as much pasture vegetation when total plant production is low and densities of sheep and goats are high. Sheep and goats may graze heavily on pasture plants, causing total intake to be higher than total production of pasture plants at very low levels of plant production. Further studies using the fecal accumulation rate technique are necessary to understand better the grazing rates of livestock.

3.3 Soil Alkalization by Overgrazing Can Delay the Recovery of Pastureland

3.3.1 *Background*

When livestock graze pasture plants heavily, plant species diversity and ground cover decline, which results in degradation of pastureland (Olff and Ritchie 1998). Therefore, most studies have focused on the intensity of grazing pressure, discussing the effects of plant-animal interactions on pasture plants (Olff and Ritchie 1998). From the perspective of plant-animal interaction, the composition of the pasture plant community should recover when livestock density decreases and grazing pressure is low. However, in Mongolian pastures, once vegetation has been degraded by overgrazing, it rarely recovers even after long periods of decreased livestock density (Fujita et al. 2012b). Thus, it is expected that an overabundance of livestock may affect pasture plants via indirect environmental changes as well as by direct overgrazing.

In addition to plant species composition, large herbivores can affect several aspects of soil condition. For example, large herbivores can cause changes in nutrient cycling (Hobbs 1996), and overgrazing can affect soil chemical properties, including carbon, nitrogen, and pH (Hiernaux et al. 1999). Reduction of ground cover increases evaporation, soil temperature, and decomposition of soil organic matter, all of which catalyze the processes of soil salinization and alkalization (Wang et al. 2009). It has also been reported that the domination of grazing-tolerant plants in pasture areas coincides with soil alkalization (Fujita et al. 2012b; Fujita in this volume). Therefore, I hypothesized that overgrazing would cause soil alkalization, and that this environmental change may delay the recovery of degraded pasture vegetation. I tested the relationships among soil alkalization, pasture plant composition, and grazing pressure using the gradient of decreasing livestock density with increasing distance from a nomadic camp. I also examined the recovery of soil pH and composition of pasture vegetation after soil alkalization due to agricultural cultivation.

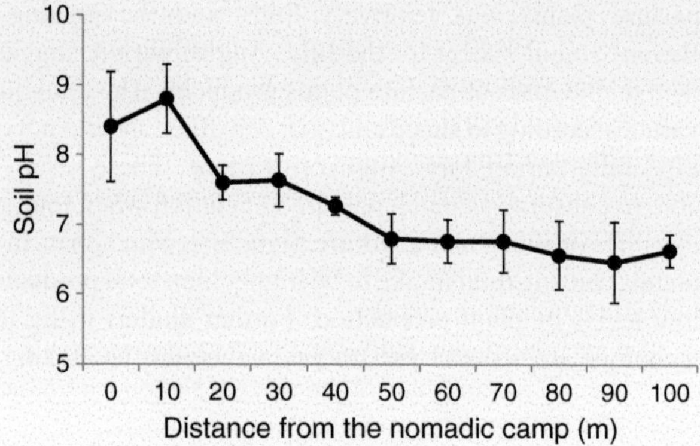

Fig. 3.10 Relationship between soil pH and distance from the nomadic camp. *Bars* indicate standard deviation

3.3.2 *Soil pH and Pasture Vegetation in Relation to Distance from a Nomadic Camp*

Three 100 × 1 m transects were established in a forest steppe area in Erdene. Each transect was extended 100 m from the nomadic camp. All nomadic camps established transects were separated by more than 900 m. Within transects, 11 1 × 1 m quadrats were established at 10-m intervals. To determine the pH of surface soil, soil from 0–2 cm was collected from each quadrat and mixed with distilled water in a small bottle, and the pH of the clear top layer of water was measured. The percentage cover and maximum height of each plant species were measured in each quadrat.

A generalized linear mixed model (GLMM) was used to assess the relationship between soil pH in each quadrat and distance from the nomadic camp, using the lme4 package of R (Bates and Maechler 2010). Difference in transect was set as a random effect, and the relationships among factors were described using a Gaussian regression model. Akaike's information criterion (AIC) was used for model selection, and models with minimum AIC values were selected as the best-fit estimator. All statistical analyses were performed using the R Language and Environment for Statistical Computing 2.11 (R Development Core Team 2010). I performed a non-metrical multidimensional scaling (NMDS) ordination to visualize the differences in composition of pasture vegetation based on the biomass index of each plant species, using the vegan package of R (Oksanen et al. 2011). Effects of soil pH and distance from the nomadic camp on composition of pasture vegetation were also tested in the NMDS. I multiplied percentage cover by maximum height for each plant species, and used the result as a biomass index.

Soil pH close to the nomadic camp was higher than that farther away from the camp (Fig. 3.10). In the GLMM analysis, the full model was selected as having minimum AIC (AIC = 73.14, AIC of null model = 92.15). Soil pH in each quadrat was predicted as follows:

$$PH = -0.021 \times DISTANCE + 8.31$$

where PH and DISTANCE indicate the soil pH in each quadrat and distance (m) from the nomadic camp, respectively. In the predicted model, soil pH near the nomadic camp was alkalized, and decreased with increasing distance from the camp ($p < 0.001$). The NMDS ordination showed clear variation in the composition of pasture vegetation. The stress value was 4.90, suggesting that the two-dimensional NMDS ordination provided a useful description of the information in the distance matrix (the stress value <20 is favorable; Legendre and Anderson 1999). Both soil pH and distance from the nomadic camp showed significant correlations with the composition of pasture vegetation ($p < 0.001$). Unpalatable annual herbs, including *Axyris amaranthoides*, *Chenopodium acuminatum*, and *Chenopodium album*, occurred frequently in the quadrat close to the nomadic camp with alkalized soil (Table 3.1; Fig. 3.11a). In contrast, palatable herbs, including *Agropyron cristatum*, *Artemisia frigida*, *Artemisia laciniata*, and *Stipa krylovii*, occurred frequently in the quadrat far from the camp, where the soil was not alkalized (Table 3.1; Fig. 3.11b). Fujita et al. (2012b) examined the relationship between soil pH and relative abundance of grazing-tolerant plants, and found that grazing-tolerant plants dominated pasture areas with alkalized soil (Fujita in this volume). My results are consistent with this report, and suggest that soil alkalization can affect the composition of pasture vegetation and allow unpalatable plants to dominate in pasture.

Livestock grazing affects soil chemical properties (Hiernaux et al. 1999). When the upward flux of groundwater with alkali elements such as Na^+ is greater than the downward flux, soil becomes salinized and alkalized (Wang et al. 2009). In turn, greater downward flux of groundwater may enhance soil acidification. In areas with high precipitation, such as northeastern Thailand and southern Queensland, land degradation causes soil acidification (Noble et al. 2000). In contrast, in areas with low precipitation, such as northeastern China, land degradation causes soil alkalization (Kawanabe et al. 1993; Wang et al. 2009). Long-term or heavy overgrazing by livestock can reduce the annual production and ground cover of pastureland (Fujita et al. 2012a). The data presented here suggest that overgrazing and trampling by large numbers of livestock can cause soil alkalization due to increased evaporation of groundwater.

3.3.3 Soil pH and Pasture Vegetation in Areas with Varying Cultivation History

To test the recovery of soil pH and pasture vegetation from heavy degradation, soil pH was measured in agricultural land located approximately 80 km west of Ulaanbaatar. I selected a continuously cultivated field, an abandoned field (around 20 years after abandonment), and a natural pasture area which had not been previously cultivated, and established one 100 × 1 m transect in each area. Eleven

Table 3.1 Total biomass indices (in 9 quadrats) of 20 main species along with the distance from the nomadic camp

	Distance from nomadic camp		
Species name	0–20 m	40–60 m	80–100 m
Agropyron cristatum	0.1	9.7	14.5
Arenaria capillaris		0.0	0.5
Artemisia dracunculus	1.7	8.7	0.8
Artemisia frigida		3.1	3.3
Artemisia laciniata		0.0	5.9
Axyris amaranthoides	2.7	0.3	0.0
Bupleurum bicaule		0.1	0.2
Calamagrostis macilenta		0.0	0.6
Carex duriuscula	0.5	10.0	4.9
Chenopodium acuminatum	62.1	2.2	
Chenopodium album	5.2	0.2	0.0
Filifolium sibiricum		0.3	1.3
Galium verum		0.3	1.9
Kochia prostrata	0.9		
Leontopodium leontopodioides		0.0	1.1
Leymus chinensis	17.6	20.7	6.2
Potentilla bifurca	0.1	0.9	0.7
Ptilotrichum canescens		0.0	0.5
Rheum undulatum	1.6		0.1
Stipa krylovii	0.0		15.1

Fig. 3.11 Conditions of pasture vegetation (**a**) close to the nomadic camp, and (**b**) far from the nomadic camp. Pasture vegetation near the nomadic camp with alkalized soils was relatively bare and consisted only of a small number of unpalatable species. Pasture vegetation far from the nomadic camp included a variety of palatable species

1 × 1 m quadrats were established at 10-m intervals within each transect. The pH of the surface soil was measured in each quadrat. In the abandoned field and natural pasture area, percentage cover and maximum height of each plant species were measured, and a biomass index was calculated for each quadrat. Differences in soil pH among the three fields were analyzed using one-way ANOVA. Tukey's HSD

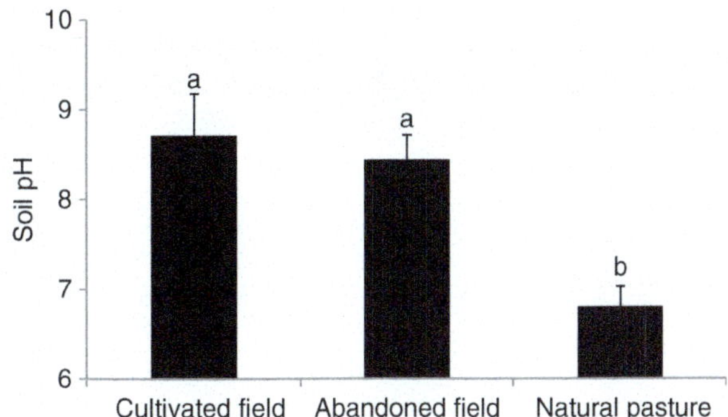

Fig. 3.12 Differences in mean soil pH among sites with different cultivation history. *Different letters* indicate significant differences (Tukey's HSD test, $p < 0.05$). *Bars* indicate standard deviation

test was employed for post-hoc analysis, significance for all tests being set at $p = 0.05$.

Soil pH differed according to cultivation history (Fig. 3.12). Soils in the cultivated and abandoned fields were strongly alkalized, and pH did not differ significantly between these fields ($p = 0.19$). Soil pH in the natural pasture area (6.8) was significantly lower than in the cultivated and abandoned fields ($p < 0.001$). The composition of pasture vegetation differed between the abandoned field and natural pastureland. Mean biomass index in the abandoned field was 39.7 and four plant species occurred in the quadrat, but only one species, *Artemisia macrocephala*, was dominant (Fig. 3.13). In natural pastureland, mean biomass index was 9.78 and 16 plant species occurred in the quadrat. *Leymus chinensis* and *Artemisia frigida*, both palatable plants, were dominant, and *A. macrocephala* was absent.

These results demonstrated that once soil is heavily alkalized by continuous cultivation, pH exhibits little recovery, even after 20 years. Sangha et al. (2005) reported that 10–15 years were necessary for soil pH to recover from alkalization after deforestation. Although the rate of recovery from soil alkalization may depend on disturbance intensity and duration, and the data presented here represent only agricultural fields, my result suggests that more than 20 years may be necessary for soils to recover from alkalization in Mongolia. Plant species diversity in pasture areas recovers quickly where soil is not alkalized, and slowly where soil has been heavily alkalized (Fujita et al. 2012b). The diversity of pasture vegetation in the abandoned field was very low, and was dominated by an unpalatable herb. Thus, it is expected that long time periods will be required for recovery of heavily alkalized soil, as found in degraded pasture area.

3.4 Conclusions

In this chapter, I have shown that a fecal accumulation rate technique can be applied to estimate livestock density at small spatial scales in Mongolia, and have used this method to estimate the grazing rate of livestock. I conclude that grazing rate is

Fig. 3.13 Vegetation in the abandoned field. The majority of the area was occupied by a single species (*A. macrocephala*)

generally constant, but that it can vary depending on total production of pasture plants. In addition, when pasture production is very low, sheep and goats may graze heavily and reduce the amount of forage available to cattle, which results in variation of the grazing rate. It is important to consider variability in the grazing rate when devising effective management strategies for livestock and pasture.

Livestock grazing also has indirect effects on soil environments. Overgrazing can cause soil alkalization, which may delay the recovery of degraded pasture vegetation. Degraded pasture areas with alkalized soil support few plant species, most of which are unpalatable for livestock. Thus, it is important to manage grazing intensity and prevent soil alkalization to achieve sustainable use of pasture.

One of the most important recent changes in government policy related to nomadic pastoralism is land privatization, which is supported by both environmentalists and developmentalists in the international community. Privatization of land has begun to occur around Mongolia's capital, resulting in a variety of urban problems (S. Kato, Chap. 6). Furthermore, land privatization may reduce the mobility of herders and lead to continuous and intense grazing, which consequently causes soil alkalization and unsustainable use of pastureland. In light of these realities, it is necessary to assess the importance of traditional nomadic pastoralism in Mongolia.

References

Bates D, Maechler M (2010) lme4: linear mixed-effects models using S4 classes. R package version 0.999375-3

Campbell D, Swanson GM, Sales J (2004) Comparing the precision and cost-effectiveness of faecal pellet group count methods. J Appl Ecol 41:1185–1196

Central Statistical Board under the Council of Ministers of the MPR (1981) National economy of the MPR for 60 years (1921–1981), Anniversary statistical collection. Ulaanbaatar, Mongolia

Côté SD, Rooney TP, Tremblay J-P, Dussault C, Waller DM (2004) Ecological impacts of deer overabundance. Annu Rev Ecol Evol Syst 35:113–147

R Development Core Team (2010) R: a language and environment for statistical computing. R Foundation for Statistical Computing, Vienna

Fernandez-Gimenez ME (2002) Spatial and social boundaries and the paradox of pastoral land tenure: a case study from post-socialist Mongolia. Human Ecol 30:49–78

Fernandez-Gimenez ME, Allen-Diaz B (1999) Testing a non-equilibrium model of rangeland vegetation dynamics in Mongolia. J Appl Ecol 36:871–885

Finch C (1999) Mongolia's wild heritage biological diversity, protected areas, and conservation in the land of Chingis Khaan. Avery Press, Boulder

Fujita N, Amartuvshin N, Yamada Y, Matsui K, Sakai S, Yamamura N (2009) Positive and negative effects of livestock grazing on plant diversity of Mongolian nomadic pasturelands along a slope with soil moisture gradient. Grassland Sci 55:126–134

Fujita N, Amartuvshin N, Ariunbold E (2012a) Annual production and species diversity of Mongolian pasture plants in relation to grazing pressure by livestock. In: Yamamura N et al (eds) The Mongolian ecosystem network: environmental issues under climate and social changes. Springer, Tokyo, pp 131–144

Fujita N, Amartuvshin N, Ariunbold E (2012b) Vegetation interactions for the better understanding of a Mongolian ecosystem network. In: Yamamura N et al (eds) The Mongolian ecosystem network: environmental issues under climate and social changes. Springer, Tokyo, pp 157–184

Gill RMA (1992) A review of damage by mammals in north temperate forests: 1. Deer. Forestry 65:145–169

Gill RMA, Beardall V (2001) The impact of deer on woodlands: the effects of browsing and seed dispersal on vegetation structure and composition. Forestry 74:209–218

Hemami MR, Watkinson AR, Dolman PM (2004) Habitat selection by sympatric muntjac (*Muntiacus reevesi*) and roe deer (*Capreolus capreolus*) in a lowland commercial pine forest. For Ecol Manage 194:49–60

Hiernaux P, Bielders CL, Valentin C, Bationo A, Fernández-Rivera S (1999) Effects of livestock grazing on physical and chemical properties of sandy soils in Sahelian rangelands. J Arid Environ 49:231–245

Hobbs NT (1996) Modification of ecosystems by ungulates. J Wildlife Manage 60:695–713

Honeychurch W (2010) Pastoral nomadic voices: a Mongolian archaeology for the future. World Archaeol 42:405–417

Hoshino A, Yoshihara Y, Sasaki T, Okayasu T, Jamsran U, Okuro T, Takeuchi K (2009) Comparison of vegetation changes along grazing gradients with different numbers of livestock. J Arid Environ 73:687–690

Kawanabe S, Oshida T, Zhu TC, Bai XK, Xing YL (1993) Degradation and conservational trial of *Aneurolepidium chinense* grassland in Northern China. Jpn J Grassland Sci 39:93–100

Koda R, Fujita N (2011) Is deer herbivory directly proportional to deer population density? Comparison of deer feeding frequencies among six forests with different deer density. For Ecol Manage 262:432–439

Koda R, Agetsuma N, Agetsuma-Yanagihara Y, Tsujino R, Fujita N (2011) A proposal of the method of deer density estimate without fecal decomposition rate: a case study of fecal accumulation rate technique in Japan. Ecol Res 26:227–231

Konagaya Y (2010) The history of agricultural development in Mongolia: seeking a tradeoff between development and conservation. Bull Nat Mus Ethnol 35:9–138

Legendre P, Anderson MJ (1999) Distance-based redundancy analysis: testing multispecies responses in multifactorial ecological experiments. Ecol Monogr 69:512

Lise W, Hess S, Purev B (2006) Pastureland degradation and poverty among herders in Mongolia: data analysis and game estimation. Ecol Econ 58:350–364

Marques FFC, Buckland ST, Goffin D, Dixon CE, Borchers DL, Mayle BA, Peace AJ (2001) Estimating deer abundance from line transect surveys of dung: sika deer in southern Scotland. J Appl Ecol 38:349–363

Melis C, Buset A, Aarrestad PA, Hanssen O, Meisingset EL, Andersen R, Moksnes A, Roskaft E (2006) Impact of red deer *Cervus elaphus* grazing on bilberry *Vaccinium myrtillus* and composition of ground beetle (*Coleoptera*, *Carabidae*) assemblage. Biodivers Conserv 15:2049–2059

MNET (2009) Mongolia assessment report on climate change 2009. Ministry of Environment, Nature and Tourism of Mongolia (MNET), Ulaanbaatar, Mongolia

Morinaga Y, Tian S, Shinoda M (2003) Winter snow anomaly and atmospheric circulation in Mongolia. Int J Climatol 23:1627–1636

Mysterud A (2006) The concept of overgrazing and its role in management of large herbivores. Wildlife Biol 12:129–141

National Statistical Office of Mongolia (2004) Mongolia in a market system, Statistical Yearbook. Ulaanbaatar, Mongolia

National Statistical Office of Mongolia (2012) Mongolian Statistical Yearbook. Ulaanbaatar, Mongolia

Noble AD, Gillman GP, Ruaysoongnern S (2000) A cation exchange index for assessing degradation of acid soil by further acidification under permanent agriculture in the tropics. Eur J Soil Sci 51:233–243

Okayasu T, Muto M, Jamsran U, Takeuchi K (2007) Spatially heterogeneous impacts on rangeland after social system change in Mongolia. Land Degrad Dev 18:555–566

Okayasu T, Okuro T, Jamsran U, Takeuchi K (2010) Impact of the spatial and temporal arrangement of pastoral use on land degradation around animal concentration points. Land Degrad Dev 21:248–259

Oksanen J, Blanchet FG, Kindt R, Legendre P, O'Hara RB, Simpson GL, Solymos P, Henry M, Stevens H, Wagner H (2011) Vegan: community ecology package. R package version 1.17-9

Olff H, Ritchie ME (1998) Effects of herbivores on grassland plant diversity. Trends Ecol Evol 13:261–265

Peters DPC (2002) Plant species dominance at a grassland – shrubland ecotone: an individual-based gap dynamics model of herbaceous and woody species. Ecol Model 152:5–32

Pogue DW, Schnell GD (2001) Effects of agriculture on habitat complexity in a prairie-forest ecotone in the Southern Great Plains of North America. Agric Ecosyst Environ 87:287–298

Reading RP, Mix HM, Lhagvasuren B, Feh C, Kane DP, Dulamtsuren S, Enkhbold S (2001) Status and distribution of khulan (*Equus hemionus*) in Mongolia. J Zool 254:381–389

Ritchie ME, Tilman D, Knops JMH (1998) Herbivore effects on plant and nitrogen dynamics in oak savanna. Ecology 79:165–177

Rooney TP, Waller DM (2003) Direct and indirect effects of white-tailed deer in forest ecosystems. For Ecol Manage 181:165–176

Rossabi M (2005) Modern Mongolia: from khans to commissars to capitalists. University of California Press, Berkeley

Saizen I, Maekawa A, Yamamura N (2010) Spatial analysis of time-series changes in livestock distribution by detection of local spatial associations in Mongolia. Appl Geogr 30:639–649

Sangha KK, Midmore DJ, Rolfe J, Jalota RK (2005) Tradeoffs between pasture production and plant diversity and soil health attributes of pasture systems of central Queensland, Australia. Agric Ecosyst Environ 111:93–103

Tremblay J-P, Huot J, Potvin F (2007) Density-related effects of deer browsing on the regeneration dynamics of boreal forests. J Appl Ecol 44:552–562

Tsogtbaatar J (2004) Deforestation and reforestation needs in Mongolia. For Ecol Manage 201:57–63

Wallis de Vris MF, Manibazar N, Dugerlham S (1996) The vegetation of the forest-steppe region of Hustain Nuruu, Mongolia. Vegetatio 122:111–127

Wang L, Seki K, Miyazaki T, Ishihama Y (2009) The causes of soil alkalinization in the Songnen Plain of Northeast China. Paddy Water Environ 7:259–270

Wingard JR, Zahler P (2006) Silent steppe: the illegal wildlife trade crisis. The World Bank 147, Washington D.C, USA

Yoshihara Y, Ito TY, Lhagvasuren B, Takatsuki S (2008) A comparison of food resources used by Mongolian gazelles and sympatric livestock in three areas in Mongolia. J Arid Environ 72:48–55

Chapter 4
Plant Diversity and Productivity of Mongolian Nomadic Pasture in Relation to Land Use

Noboru Fujita and Erdenegerel Ariunbold

Abstract About 40 % of the population of Mongolia is involved in nomadic pastoralism. Thus, maintaining productive pasture through regulation is an important topic of research. To gain a better understanding of the ecological interactions between pasture plants and livestock, we conducted a field study of Mongolian pasturelands. We found that intermediate disturbance by livestock grazing maximized both species richness and annual productivity of pasture plants, although the optimal intensity of grazing differs with soil moisture and fecundity. However, too intensive continuous grazing decreases plant species richness and aboveground annual production within several years. Intensive grazing may lead to nearly irreversible changes associated with dominance of grazing-tolerant plants unpalatable to animals and soil alkalization. Pasture degradation is often attributed to a recent increase in goats kept for cashmere production. We confirmed some difference in food preferences between goats and sheep, but both can cause severe damage to pasture plants.

Keywords Goats • Nomadic pastoralism • Pasture degradation • Sheep • Soil alkalization

4.1 Introduction

Nomadic pastoralism is a form of pastoralism in which livestock are herded to fresh pastures on which to graze following an irregular pattern of movement. Nomadic pastoralism is not a relict but a primary industry over much of Mongolia, although

N. Fujita (✉)
Research Institute for Humanity and Nature, Kyoto, Japan
e-mail: fujitanb@ae.auone-net.jp

E. Ariunbold
Institute of Geoecology, Mongolian Academy of Science, Ulan Bator, Mongolia

S. Sakai and C. Umetsu (eds.), *Social–Ecological Systems in Transition*, Global Environmental Studies, DOI 10.1007/978-4-431-54910-9_4, © Springer Japan 2014

its relative importance is declining due to recent rapid growth of the mining industry. About 40 % of the population is involved in nomadic pastoralism, supplying livestock products throughout Mongolia as well as to the international market.

Under nomadic pastoralism, the grazing patterns of the livestock greatly affect the condition of the pasture, which is a key factor in sustainable, productive, profitable, and healthy livestock production. Traditional knowledge of how to manage effectively livestock grazing of the pasture has been accumulated by herders in Mongolia over centuries. Maintaining productive pasture through formal and informal regulation is an important research topic in Mongolia, particularly because of recent rapid changes in the practice of nomadic pastoralism due to substantial structural changes in the social and economic systems of the country (Koda in this volume).

In this chapter we discuss relationships between the practice of nomadic pastoralism (grazing pressure) and the condition of the pasture from an ecological perspective, which may be useful in managing productive and sustainable use of the Mongolian pasture. In this study, grazing pressure has been represented by the frequency of grazing and the height of the remaining vegetation after grazing. We have evaluated pasture condition using species richness and annual productivity.

4.2 Relationships Between Grazing Pressure, Plant Species Richness, and Soil pH

In tropical rain forests and coral reefs, Connell (1978) showed that an intermediate level of disturbance maximized species richness. In pasture, the intermediate disturbance hypothesis has rarely been studied (but see Hiernaux 1998). For pasture, livestock grazing can be regarded as a disturbance, and thus a method for measuring the degree of disturbance intensity is needed. The height of pasture plants at a given time depends on both the grazed height and the grazing interval of livestock. The height of pasture plants is lower with a greater grazing pressure above the ground when the frequency of grazing is equal. When livestock always graze at the same height, the height of the pasture plants becomes lower with more frequent grazing due to shorter recovery time after grazing. Therefore, plant height may serve as an index of grazing pressure.

At a flat pasture along a river in the forest steppe zone near Ihtamir, Arkhangai, we observed the relationship between the height and species richness of the plants. Species richness was evaluated by counting the number of plant species within a 1-m^2 plot in the pasture. Species richness was highest with intermediate grazing pressure (Fig. 4.1), supporting the intermediate disturbance hypothesis. A decrease in plant species richness with decreasing grazed plant height occurred near the herder's tent and winter shed, where the grazing pressure by livestock tends to be very high due to the gathering of the sheep, goats, and cattle there each night.

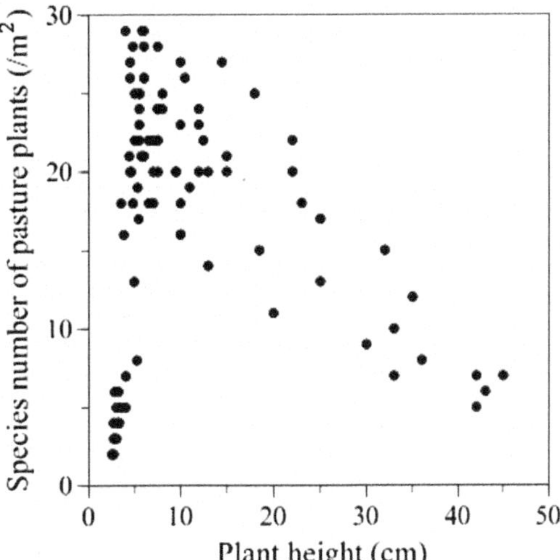

Fig. 4.1 Relationship between plant height and plant species richness (per square meter) in pastures (Fujita et al. 2002)

A decrease in plant species richness with increased plant height, on the other hand, was found at sites where grazing-tolerant plants were dominant with infrequent grazing. At this site, the pasture had been degraded by repeated use of the same pasture.

Relationships between plant height, soil pH, and percent coverage of grazing-tolerant plants in a 1-m^2 plot of pasture (Fig. 4.2) show that low percent coverage of grazing-tolerant plants occurred at sites where plant height was intermediate or low and the soil was weakly acid or neutral. A high percent coverage of grazing-tolerant plants was found in plots where plant height and soil pH was high. When the winter shed was located in a pasture for only one winter, the plant species richness decreased rapidly to two or three species, but recovered 4 years after moving the winter shed site, because short-term overgrazing does not cause the soil pH to become alkaline. However, recovery of species richness is very slow once pastures become alkaline (Fujita et al. 2012b).

The process of pasture degradation along with a decline in plant species richness can be summarized as follows. Initially, pasture grazed at intermediate grazing pressure by livestock maintains high plant species richness and is dominated by plants edible by livestock with weakly acid soil pH. When the pasture is overgrazed, the plant species richness falls, but the soil pH does not become alkaline immediately. However, if overgrazing continues for a long time, the pasture becomes dominated by grazing-tolerant plants with soil alkalization. Finally, dominance of grazing-tolerant plants continues indefinitely once a pasture becomes alkaline despite decreased grazing pressure. In contrast, plant species richness may recover quickly when the soil has not become alkaline once overgrazing is discontinued (Fujita et al. 2012b).

Overgrazing near the herder's tent and winter shed is inevitable in nomadic pastoralism, because livestock are gathered there every day. However, a short-term

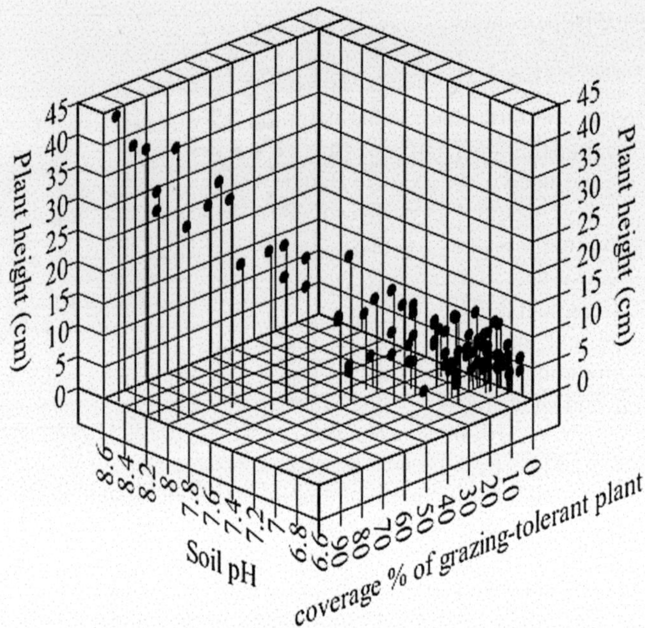

Fig. 4.2 Relationships between plant height, soil pH, and percent coverage of grazing-tolerant plants of the total coverage of all plants in the pasture

use, for instance, one season, does not cause permanent pasture degradation. Continuous overgrazing through many seasons without migration causes pasture degradation. For pasture productivity, it has been reported that rotational grazing management results in high pasture productivity compared with continuous grazing management (Paine et al. 1999).

4.3 Effects of Habitat Fecundity on Plant Species Richness in Grazed Pasture

Although we found the highest plant species richness in pasture with intermediate grazing pressure, relationships between species richness and grazing pressure have not been consistent between studies. Some studies have reported positive correlations (Collins et al. 2002; Bakker et al. 2003; Frank 2005), while others reported negative (Wilsey and Polley 2003; Guo 2004; Hendricks et al. 2005) or no correlation (Tracy and Sanderson 2000). Proulx and Mazmunder (1998) suggested that these varying results were due to different environmental conditions, and hypothesized that plant species richness increases with increased grazing pressure in nutrient-rich environments, while it decreases in nutrient-poor environments.

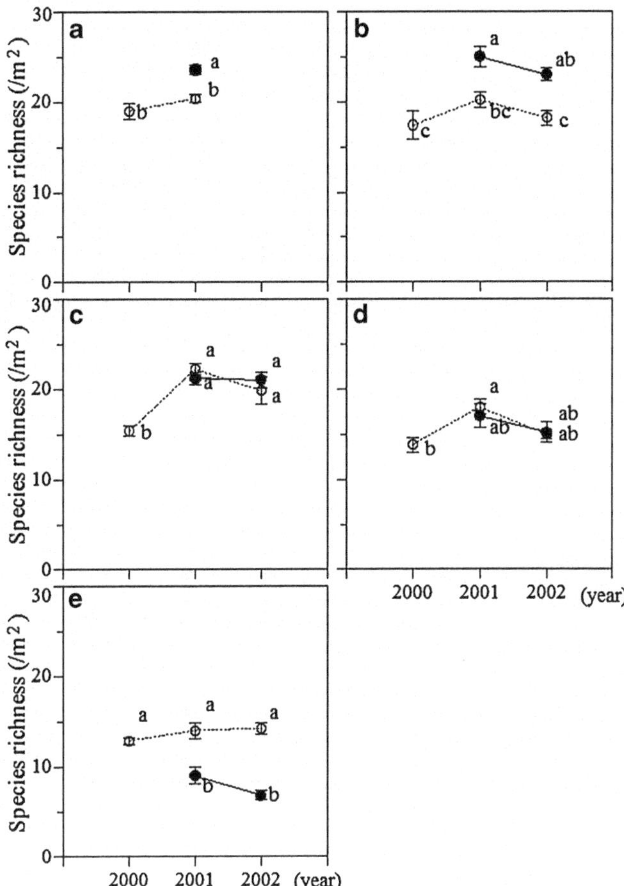

Fig. 4.3 Plant species richness inside and outside an exclosure (**a**) near the ridge, (**b**) upper slope, (**c**) lower slope, (**d**) foot of the slope, and (**e**) flat valley bottom. *White circles*: outside the exclosure; *black circles*: inside the exclosure. *a*, *b*, and *c* on the *bar* and *line* indicate statistical significance ($P < 0.01$) (Fujita et al. 2009)

To compare the effects of grazing on species richness under different nutrient conditions, a ridge-valley gradient on a slope in a Mongolian nomadic pasture is an ideal system. Soil moisture increases along a downward slope in just a short distance (Yanagisawa and Fujita 1999) and nutrient availability typically increases with soil moisture. Grazing pressure will be similar throughout the slope because sheep and goats, the two dominant livestock animals in Mongolia, tend to graze without distinguishing between valleys, slopes, or ridges.

To compare species richness in pastures with and without grazing under different habitat fecundity conditions, we constructed exclosures near the ridge (Fig. 4.3a), on the upper slope (Fig. 4.3b), on the lower slope (Fig. 4.3c), at the foot of the slope (Fig. 4.3d), and on the flat valley bottom (Fig. 4.3e) along a slope with pasture in the forest steppe zone near Ulan Bator. In the exclosures, pastures were protected from grazing, while the pastures outside were freely accessed by animals. The following summer we found that plant species richness was higher in the exclosures than outside them near the ridge and on the upper slope, while the difference was minimal on the lower slope and at the foot of the slope, and species richness was lower in the exclosure than outside it on the flat valley bottom (Fig. 4.3). Therefore, the hypothesis of Proulx and Mazmunder (1998) was supported by our data.

These results can be explained by the relative strength of the exclusion effect on plants by livestock grazing vs the reduction in plant competition for light by livestock grazing. Near the ridge and on the upper slope where plant growth was low due to dry soil, the exclusion effect on plants by livestock grazing lowered plant species richness despite low plant competition for light. On the other hand, on the flat valley bottom where plant growth was high due to moist soil, plant species richness decreased in the exclosure because competition was stronger, reducing diversity. This result is consistent with previous studies showing that plant species diversity decreases with habitat fertility (Wilson and Tilman 1991; Campbell and Grime 1992; Rajaniemi 2002) due to greater competitive exclusion between plants in habitats with higher productivity (Collins et al. 1998; Rajaniemi 2003). Although good nutrient supply in nutrient-rich habitats eases plant competition for nutrients (Huston and DeAngelis 1994), stronger growth leads to more intense competition for light. Cattle grazing increases the species richness of small- and medium-sized pasture plants, while it decreases the species richness of tall pasture plants (Pykälä 2004). Livestock grazing that reduces plant competition for light between pasture plants may assure survival of less competitive plants and increase plant species richness in nutrient-rich habitats. On the lower slope and at the foot of the slope, plant species richness was similar inside and outside the exclosures, because the exclusion effects and competition reducing effects were balanced.

The results of our experiments demonstrate that habitat fecundity affects plant species richness in pastures with livestock grazing by regulating the plant growth rate and the intensity of competition for light between plants. In addition, the intensity of grazing pressure that results in the highest plant species richness may increase with increased habitat fecundity or plant growth rate. In Mongolia, along with grazing pressure, the maximum species diversity of pasture plants was found in the most heavily grazed plots in the forest steppe zone, in the least-grazed plots in the steppe zone, and outside the grazed plots in the desert steppe zone (Fernandez-Gimenez and Allen-Diaz 1999). This pattern generally agrees with the hypothesis above.

4.4 Results for Different Grazing Intensities

Although degradation of pastures is usually attributed to overgrazing, it is often unclear specifically how strong grazing causes degradation. Our studies showed that monthly 3-cm-high mowing of pasture plants resulted in the maximum annual productivity of pasture plants from different treatments with various frequencies of mowing and grazing heights (Fujita et al. 2012a). Therefore, grazing by livestock with equivalent effects to mowing more frequent than monthly and lower than 3 cm can be regarded as "overgrazing."

To determine the effects of grazing at different intensities on species richness and annual productivity of pasture plants, we conducted experimental mowing at various frequencies and heights of pasture plants in exclosures for several years. We selected monthly and half-monthly mowing for frequency, and 3 cm and 0 cm for mowing height. Each experiment was conducted in a quadrat of 1 m^2 in an exclosure, and had five replicates. The study sites were flat pastures at Gachuurt (48°00′74″ N, 107°11′26″ E) in the forest steppe zone, Mandalgobi (45°43′44″ N, 106°16′16″ E) in the steppe zone, and Dalanzadgad (43°34′55″ N, 104°25′40″ E) in the dry steppe zone, and were evaluated for 4 years from 2006 to 2009. In addition, monthly mowing at 0- and 3-cm heights was conducted for 3 years at Erdene (47°39′29″ N, 107°43′58″ E) in the forest steppe zone and Bayan-Unjuul (47°03′46″ N, 105°44′48″ E) in the steppe zone from 2009 to 2011. We placed the exclosures in autumn the year before the first experimental year, and mowed from late May/early June to mid-late August during the growing season each year. At control plots we mowed once to a 3-cm height in mid-August at all sites. These experiments were continued every year in the same quadrat to examine the long-term effects of different grazing pressures.

The dry weight of the trimmings was measured after each mowing after drying at 80 °C for 2 days. Just prior to the last mowing in mid-August, we recorded the plant species to determine the species richness of the pasture plants in each quadrat. We dug up the underground parts of the plants in five quadrats for the 3-cm mowing height and monthly mowing interval, the 0-cm mowing height and half-monthly mowing interval, and the control in early October 2009 (Gachuurt) or early May 2010 (Mandalgobi and Dalanzadgad) after the growing season of 2009 to determine the differences in underground biomass between the treatments after 4 years. At Erdene and Bayan-Unjuul, we dug up the underground parts of the plants in five quadrats for the 3-cm and 0-cm mowing heights with monthly mowing interval and the control in early October 2011 after the growing season of 2011. After washing the rhizomes with buds and the roots, we measured the total dry weight in each quadrat in the same manner as the aboveground parts.

The annual productivity and species diversity of pasture plants for each mowing experiment are shown in Fig. 4.4 (Gachuurt, Mandalgobi, and Dalanzadgad) and Fig. 4.5 (Erdene and Bayan-Unjuul). Annual variations in species richness and productivity of pasture plants were due to fluctuations in precipitation. For instance, low species richness and low annual productivity at Gachuurt and Mandalgobi in 2007 were caused by low annual precipitation. The timing of precipitation also affected the aboveground biomass, particularly for the control plots, which were mowed only once in August. Very low aboveground annual productivity of the control in 2008 at Dalanzadgad was caused by very low precipitation in August, which killed most aboveground plants that had already grown.

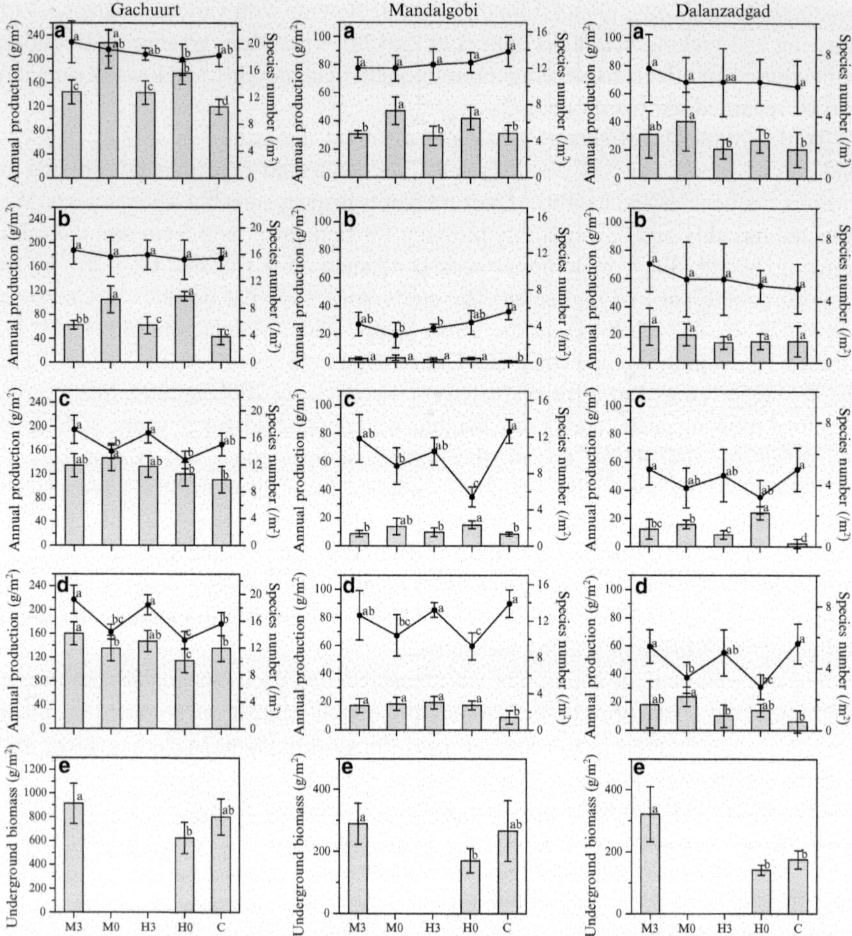

Fig. 4.4 Annual production, species diversity, and underground biomass of pasture plants with different mowing treatments at Gachuurt, Mandalgobi, and Dalanzadgad: (**a**) the first year of mowing in 2006, (**b**) the second year of mowing in 2007, (**c**) the third year of mowing in 2008, (**d**) the fourth year of mowing in 2009, and (**e**) underground biomass of pasture plants after 4 years. *M3* monthly mowing to a 3-cm plant height, *M0* monthly mowing to a 0-cm plant height, *H3* mowing at half-monthly intervals to a 3-cm plant height, *H0* mowing at half-monthly intervals to a 0-cm plant height, *C* a single mowing event in August to a 3-cm plant height. *a*, *b*, *c*, and *d* on the *bar* and *line* show statistical significance ($P < 0.01$)

At all sites, in the first mowing year, the species richness of pasture plants did not differ between the treatments, while the aboveground annual productivity was higher for the 0-cm mowing than for the 3-cm mowing. However, after mowing in successive years, the species richness was lower for the 0-cm mowing than for the 3-cm mowing at all sites, and differences in the aboveground annual productivity between the 0-cm and 3-cm mowing decreased and were not significant

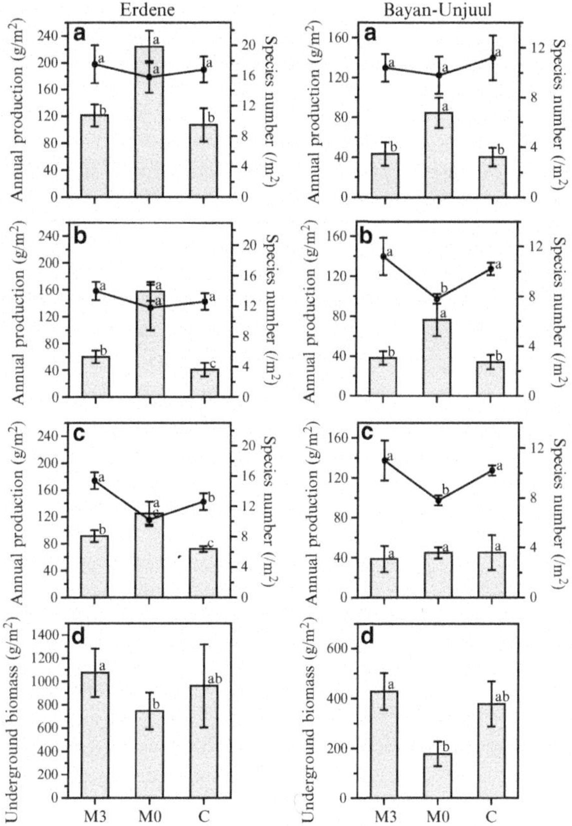

Fig. 4.5 Annual production, species diversity, and underground biomass of pasture plants mowed to 3-cm height at different intervals in Erdene and Bayan-Unjuul: (**a**) the first year of mowing in 2009, (**b**) the second year of mowing in 2010, (**c**) the third year of mowing in 2011, and (**d**) underground biomass of pasture plants after 3 years. M3, M0, C, *a*, *b*, and *c* are the same as in Fig. 4.4

except at the Erdene site. Moreover, at Gachuurt, the aboveground annual productivity of the 0-cm mowing became lower than that of the 3-cm mowing; thus, the pattern observed in the first year had reversed by the final year.

Higher aboveground annual productivity of the 0-cm mowing in the first year may not have been due to improved efficiency of photosynthesis, but rather to compensatory growth (McNaughton 1983) as a result of translocation from underground storage. At the end of the experiment, we examined the underground biomass of the plants in some treatments. The underground biomass for the 0-cm mowing was less than for the 3-cm mowing at all sites. The biomass was slightly higher for the 3-cm mowing than for the control at almost all sites, although the difference was rarely significant. Maximization of annual productivity by intermediated grazing pressure can occur through an increase in photosynthesis due to improved photosynthetic efficiency rather than compensatory growth, and strong aboveground growth results in greater underground storage (Fujita et al. 2012a). Although maximized productivity with intermediate grazing pressure is often attributed to nutrient cycling (McNaughton 1979; Hilbert et al. 1981;

Dyer et al. 1986), Fujita et al. (2012a) reported that intermediate grazing pressure optimized the annual productivity of pasture even without nutrient recycling.

Species richness was higher for the monthly 3-cm mowing than for the control at Gachuurt and Erdene, located in the forest steppe zone where nutrient conditions are generally better than in the steppe or dry steppe zones. As noted above, the effects of grazing pressure on species richness depends on the nutrient conditions of the sites (Proulx and Mazmunder 1998; Fujita et al. 2009). The observed differences between the sites are reasonably consistent with the hypothesis examined in the previous section.

The results of this experiment suggest that repeated grazing of most aboveground biomass by livestock for several successive years can be regarded as "overgrazing" causing pasture degradation by decreasing regrowth ability of the pasture plants. Underground biomass cannot be measured without digging up and killing the plant, but reduction in the plant species richness is a good indicator of pasture degradation due to overgrazing. The effect of grazing height may be more pronounced than that of grazing frequency, because monthly mowing at 0-cm height was followed by the compensatory growth while 5 days' interval of mowing at 3-cm height was not (Fujita et al. 2012a).

Evaluation of carrying capacity or appropriate density of animals of the pasture is very difficult (McLeod 1997). If we define carrying capacity as the density of animals sustained in a pasture, keeping livestock with density that exceeds the carrying capacity for a long time should be impossible by definition. However, the compensatory growth of pasture plants make it possible to keep more animals than carrying capacity for a limited time period, i.e., for at least several years. This is one reason why overgrazing could occur even though no herder deliberately wants to degrade the pasture.

Degraded pastures distinguished by low species richness and less annual production can recover relatively easily if herders avoid using pastures with less food for animals. However, extremely high grazing pressure can lead to more serious pasture degradation characterized by domination by grazing-tolerant plants, which have various mechanisms to defend themselves against herbivores (Levin 1976; Crawley 1983). In such pastures, soil alkalization is also frequently observed (Fujita et al. 2012b). Since grazing-tolerant plants grow well and most other plant species cannot grow on the alkalized soil, soil alkalization maintains domination by grazing-tolerant plants and retards the recovery of the pasture for herding for a long time (Koda in this volume). Therefore, herders should migrate before serious pasture degradation occurs (Kato in this volume); migration only after serious degradation corresponds to the tragedy of the commons (Hardin 1968).

4.5 Are Goats or Sheep Responsible for Pasture Degradation?

There have been many reports on the differences in dietary preferences between sheep and goats. Goats consume more trees and shrubs than do sheep (Walker et al. 1994; Magadlela et al. 1995; Bartolomé et al. 1998; Rogosic et al. 2006; Sanon et al. 2007). Sheep and goats prefer more nutritious grasses and forbs (Bartolomé et al. 1998; Animut et al. 2005), and can be classified into grazers and intermediate between grazers and browsers (Hofmann 1989), respectively. These results have been obtained from longtime field observations or experiments, but these studies often do not clearly distinguish cause and effect. To understand clearly the effects of overgrazing by goats and sheep on vegetation, we conducted a short-term experiment at a shrub site and a herbaceous site in the steppe zone in Bayan-Unjuul (Fujita et al. 2012b).

The results of our study generally supported the previous results. In August 2010, we raised goats and sheep in small enclosures for several days at a shrub site (Fig. 4.6). The goats preferred shrub (*Caragana*) leaves, while the sheep preferred herbs. Four days after the start of experiment, the goats maintained their weight in the *Caragana*-dominated exclosure due to large biomass of *Caragana* leaves, while the sheep lost weight in all exclosures because of the shortage of herbs. After 6 days, the shrub leaves and perennials were nearly fully grazed by the hungry goats and sheep. In August 2011 we conducted a similar experiment at the herbaceous site. Three days after the start of the experiment, the goats maintained their weight because they grazed on all herbs equally, while the sheep lost weight. The lower the

Fig. 4.6 Vegetation just before the experiment with the exclosure in August 2010; the shrub species *Caragana microphylla* and perennials such as *Stipa krylovii* and *Artemisia frigida* were dominant

Fig. 4.7 Recovered vegetation in August 2011 after the 2010 experiment with the exclosure. Leaves of *Caragana microphylla* recovered well; however, perennials whose roots were grazed did not recover. The yellow flower is *Artemisia annua*, an annual which grew after removal of the perennials

grass (*Stipa*) coverage, the more weight the sheep lost due to their special preference for *Stipa*.

At the shrub site, 6 days after the start of experiment, both the goats and sheep suffered from strong hunger and grazed not only on the aboveground parts of the herbs but on the underground parts in some areas of the exclosure. We removed the goats and sheep from the exclosure on the 6th day and maintained the exclosure until the next summer. At that time, the shrub leaves had recovered well, but the perennials were unable to recover in the areas where the underground parts of the plants had been grazed (Fig. 4.7). Both the goats and the sheep grazed the underground parts of the perennial herbs when very hungry (Fig. 4.8), and the damage to the perennial herbs was more severe when the density of the animals was higher (Fig. 4.9).

Although our experiments showed some differences in feeding preferences between goats and sheep, we consider animal density to be the most critical factor for pasture degradation rather than the animal species. Goats are often accused of causing pasture degradation due to hard grazing down to the underground parts of pasture plants compared to sheep, because a high density of goats are found in desertified areas (Kobayashi et al. 2005). However, goats prefer shrubs while sheep prefer herbs. The sheep did not affect the mature shrubs, while heavy goat pressure destroyed several shrubs (Harrington 1979). Although *Caragana* recovered in the following season in our experiments, heavier and more continuous goat grazing

Fig. 4.8 Vegetation in the exclosure after 3 days with (**a**) sheep, somewhat strong grazing, and (**b**) goats, generally grazing

might degrade palatable shrub vegetation. Our results, however, showed that goats and sheep did not differ in degrading pastures by grazing the underground parts of perennials when hungry, and that livestock density was the main factor in degrading perennial herbs. For pasture conservation, overpopulation of any kind of livestock

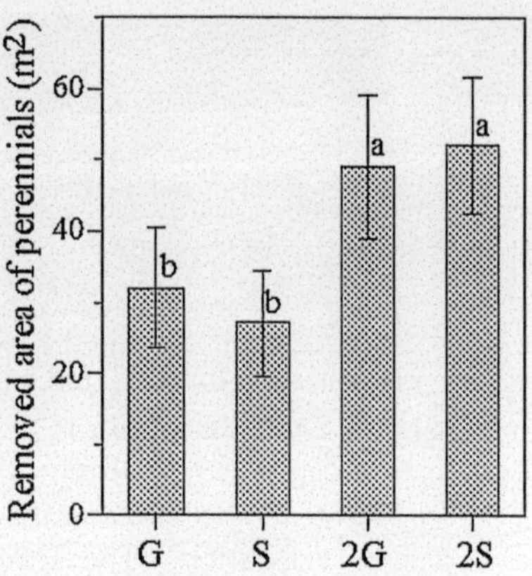

Fig. 4.9 Decreased area of perennials due to livestock grazing of roots over 6 days within a 10 × 10 m^2 fence. *G* one goat, *S* one sheep, *2G* two goats, *2S* two sheep

should be avoided. Overgrazing in some areas is caused not only by overpopulation of livestock, but by continuous intensive grazing due to settlement of livestock. Migration of livestock as in pastoral nomadism in Mongolia is effective for avoiding continuous grazing.

Which is better adapted to arid habitat, goats or sheep? Generally, goats are raised in arid lands more than are sheep. There have been several reports that goats have lower water consumption needs than sheep (Dawson et al. 1975; Gihad 1976; Ferreira et al. 2002, but see McGregor 1975). Another reason may be diet preference. Generally, in arid lands there is a dominance of shrubs over herbs, consistent with the food preferences of goats. However, our results showed there were no dietary differences between goats and sheep under very hungry conditions. Goats are highly adaptable in their diet, while sheep are less so and often experience weight loss depending on the plant species composition. They lost weight easily when there was a shortage of their preferred food such as *Stipa* (Poaceae), even if a sufficient amount of vegetation was available (Fujita et al. 2012b). Thus, goats are easier to raise in arid lands than sheep, because the diet preferred by sheep is not abundant in arid lands.

4.6 Conclusions

The highest species richness and highest annual productivity of pasture plants can be realized in pasture systems with interaction between pasture plants and livestock. Both species richness and annual productivity of pasture plants were highest with some livestock grazing, although the intensity of grazing that resulted in the highest diversity and productivity differed depending on the soil fertility. These results suggest that herding is an effective way to manage Mongolian pastures. However, pasture degradation could occur through inappropriate practices, as observed in our study. For example, too intensive or frequent grazing for several or more consecutive years decreases underground storage of pasture plants, resulting in a reduction in species richness and annual productivity. Although annual productivity may increase for the first few years of intensive grazing, this increase is not due to increased photosynthetic gain but to compensatory growth supported by underground storage. Compensatory growth enables livestock populations to be maintained above the carrying capacity for several years; however, adoption of measures to prevent pasture degradation should not be delayed. Under more intensive grazing, phenomena such as dominance by grazing-tolerant plants and soil alkalization occurs (Fujita et al. 2012b).

Dietary preferences are different between goats and sheep, which prefer shrub leaves and herbs, respectively. However, when very hungry, both goats and sheep degrade perennial pasture similarly by grazing on the underground parts of perennials. This short-term degradation of perennial pasture due to underground grazing by hungry livestock is not typical of Mongolian nomadic pastoralism. Thus, appropriate animal density rather than composition is essential for maintaining productive pastures, although pasture degradation is often attributed to the recent increase in keeping goats for cashmere production. It is necessary to continue to accumulate ecological knowledge on plant-livestock interactions for nomadic herding, not only for sustainability but also to adapt flexibly to changing social needs.

References

Animut G, Goetsch AL, Aiken GE, Puchala R, Detweiler G, Krehbiel CR, Merker RC, Sahlu T, Dawson LI, Johnson ZB, Gipson TA (2005) Performance and forage selectivity by sheep and goats co-grazing grass/forb pastures at three stocking rates. Small Rumin Res 59:203–215

Bakker C, Blair JM, Knapp AK (2003) Does resource availability, resource heterogeneity or species turnover mediate changes in plant species richness in grazed grasslands? Oecologia 137:385–391

Bartolomé J, Franch J, Plaixat J, Selingman NG (1998) Diet selection by sheep and goats on Mediterranean heath-woodland range. J Range Manage 51:383–391

Campbell BD, Grime JP (1992) An experimental test of plant strategy theory. Ecology 73:15–29

Collins SL, Knapp AK, Briggs JM, Blair JM, Steinauer EM (1998) Modulation of diversity by grazing and mowing in native tallgrass prairie. Science 280:745–747

Collins SL, Glenn SM, Briggs JM (2002) Effect of local and regional processes on plant species richness in tallgrass prairie. Oikos 99:571–579

Connell JH (1978) Diversity in tropical rain forests and coral reefs. Science 199:1302–1310

Crawley MJ (1983) Herbivory: the dynamics of animal-plant interactions. University of California Press, Berkeley

Dawson TJ, Denny MJS, Russell EM, Ellis B (1975) Water usage and diet preferences of free ranging kangaroos, sheep and feral goats in the Australian arid zone during summer. J Zool 177:1–23

Dyer MI, DeAngelis DL, Post WM (1986) A model of herbivore feedback on plant productivity. Math Biosci 79:171–184

Fernandez-Gimenez ME, Allen-Diaz B (1999) Testing a non-equilibrium model of rangeland vegetation of dynamics in Mongolia. J Appl Ecol 36:871–885

Ferreira AV, Hoffman LC, Schoeman SJ, Sheridan R (2002) Water intake of Boer goats and mutton merinos receiving either a low or high energy feedlot diet. Small Rumin Res 43:245–248

Frank DA (2005) The interactive effects of grazing ungulates and aboveground production on grassland diversity. Oecologia 143:629–634

Fujita N, Amartuvshin N, Uchida T, Wada E (2002) Biodiversity and sustainability of Mongolian herbaceous plants subjected to nomadic grazing. In: Fujita N, Timoshkin OA, Urabe J, Wada E (eds) Sustainable watersheds with emphasis on lake ecosystems, DIWPA Series, vol 3. Nauka-Center, Novosibirsk

Fujita N, Amartuvshin N, Yamada Y, Matsui K, Sakai S, Yamamura N (2009) Positive and negative effects of livestock grazing on plant diversity of Mongolian nomadic pasturelands along a slope with soil moisture gradient. Grass Sci 55:126–134

Fujita N, Amartuvshin N, Ariunbold E (2012a) Annual production and species diversity of Mongolian pasture plants in relation to grazing pressure by livestock. In: Yamamura N, Fujita N, Maekawa A (eds) The Mongolian ecosystem network. Springer, Tokyo

Fujita N, Amartuvshin N, Ariunbold E (2012b) Vegetation interactions for the better understanding of a Mongolian ecosystem network. In: Yamamura N, Fujita N, Maekawa A (eds) The Mongolian ecosystem network. Springer, Tokyo

Gihad EA (1976) Intake, digestibility and nitrogen utilization of tropical natural grass hay by goats and sheep. J Anim Sci 43:879–883

Guo Q (2004) Slow recovery in desert perennial vegetation following prolonged human disturbance. J Veg Sci 15:757–762

Hardin G (1968) The tragedy of the commons. Science 162:1243–1248

Harrington G (1979) The effects of feral goats and sheep on the shrub populations in a semi-arid woodland. Aust Rangel J 1:334–345

Hendricks HH, Bond WJ, Midgley JJ, Novellie PA (2005) Plant species richness and composition a long livestock grazing intensity gradients in a Namaqualand (South Africa) protected area. Plant Ecol 176:19–33

Hiernaux P (1998) Effects of grazing on plant species composition and spatial distribution in rangelands of the Sahel. Plant Ecol 138:191–202

Hilbert DW, Swift DM, Detling JK, Coppock DL (1981) Relative growth rates and the grazing optimization hypothesis. Oecologia 51:14–18

Hofmann RR (1989) Evolutionary steps of ecophysiological adaptation and diversification of ruminants: a comparative view of their digestive system. Oecologia 78:443–457

Huston MA, DeAngelis DL (1994) Competition and coexistence: the effects of resource transport and supply rates. Am Nat 144:954–977

Kobayashi T, Nakayama S, Wang L, Li G, Yang J (2005) Socio-ecological analysis of desertification in the Mu-Us Sandy Land with satellite remote sensing. Landsc Ecol Eng 1:17–24

Levin DA (1976) The chemical defenses of plants to pathogens and herbivores. Ann Rev Ecol Sys 7:121–159

Magadlela AM, Dabaan ME, Bryan WB, Prigge EC, Skousen JG, D'Souza GE, Arbogast BL, Flores G (1995) Brush clearing on hill land pasture with sheep and goats. J Agron Crop Sci 174:1–8

McGregor BA (1975) Water intake of grazing Angora wether goats and Merion wether sheep. Austra J Exp Agri 26:639–642

McLeod SR (1997) Is the concept of carrying capacity useful in variable environments? Oikos 79:529–542

McNaughton SJ (1979) Grazing as an optimization process: grass-ungulate relationships in the Serengeti. Am Nat 113:691–703

McNaughton SJ (1983) Compensatory plant growth as a response to herbivory. Oikos 40:329–336

Paine LK, Undersander D, Casier MD (1999) Pasture growth, production, and quality under rotational and continuous grazing management. J Prod Agri 12:569–577

Proulx M, Mazmunder A (1998) Reversal of grazing impact on plant species richness in nutrient-poor vs. nutrient-rich ecosystems. Ecology 79:2581–2592

Pykälä J (2004) Cattle grazing increases plant species richness of most species trait groups in mesic semi-natural grasslands. Plant Ecol 175:217–226

Rajaniemi TK (2002) Why does fertilization reduce plant species diversity? Testing three competition-based hypotheses. J Ecol 90:316–324
Rajaniemi TK (2003) Explaining productivity-diversity relationships in plants. Oikos 101:449–457
Rogosic J, Pfister JA, Provenza FD, Grvesa D (2006) Sheep and goat preference for and nutritional value of Mediterranean maquis shrubs. Small Rumin Res 64:169–179
Sanon HO, Kabore-Zoungrana C, Ledin I (2007) Behaviour of goats, sheep and cattle and their selection of browse species on natural pasture in a Sahelian area. Small Rumin Res 67:64–67
Tracy BF, Sanderson MA (2000) Patterns of plant species richness in pasture lands of the northeast United States. Plant Ecol 149:169–180
Walker JW, Kronberg SL, Al-Rowaily SL, West NE (1994) Comparison of sheep and goat preferences for leafy spurge. J Range Manage 47:429–434
Wilsey BJ, Polley HW (2003) Effects of seed additions and grazing history on diversity and productivity of subhumid grasslands. Ecology 84:920–931
Wilson SD, Tilman D (1991) Components of plant competition along an experimental gradient of nitrogen availability. Ecology 72:1050–1065
Yanagisawa N, Fujita N (1999) Different distribution patterns of woody species on a slope in relation to vertical root distribution and dynamics of soil moisture profiles. Ecol Res 14:165–177

Chapter 5
Reshaping Neighborhood Parks for Biodiversity and People: A Case of Unsung Socio-Ecological Systems in Bangalore, India

Savitha Swamy and M. Soubadra Devy

Abstract Urban green spaces have recently gained a lot of attention, as they are known to provide various vital ecosystem services to the community. Bangalore, a south-Indian city, which was called the "Garden City" of India, has several large green spaces. It is only in recent years that small pocket green spaces such as neighborhood parks have been created. Although the importance of neighborhood parks is known, they are ignored and readily sacrificed for developmental projects, while the large heritage green spaces receive more attention and are conserved. The concept of the large spaces providing more services seems to have filtered into the minds of citizens, thus resulting in complete negligence towards the neighborhood parks. Cities are required to implement newer concepts which focus on small green spaces too, which could enhance the services they currently provide to the community. Thus integrating multiple concepts that not only focus on the ecological functioning but also the social needs of the community could help increase the stewardship which is currently lacking around neighborhood parks and much needed attention towards small green spaces. In this chapter, through an interdisciplinary approach, we suggest concepts that could help conserve smaller green spaces through better green space management in developing cities.

Keywords Bangalore • Ecosystem services • Neighborhood parks • Reconciliation ecology • Socio-ecological systems

---

S. Swamy (✉) • M.S. Devy
Ashoka Trust for Research in Ecology and the Environment (ATREE),
Royal Enclave, Srirampura, Jakkur Post, Bangalore 560064, India
e-mail: savitha.swamy7@gmail.com

S. Sakai and C. Umetsu (eds.), *Social–Ecological Systems in Transition*, Global Environmental Studies, DOI 10.1007/978-4-431-54910-9_5,

5.1 Introduction

Ecosystems the world over are under tremendous pressure and the scope and nature of their modification has changed drastically (Vitousek et al. 1997). Although modification of natural ecosystems cannot be avoided, because they satisfy basic human requirements such as food and shelter, there is an urgent need to conserve and achieve sustainability of the services and resources that ecosystems provide (Kenward et al. 2011). Cities have always drawn on their surrounding ecosystems for goods and services (Folke et al. 1997; Rees 1997; Rees 2003). Over the last decade, rapid development has resulted in two distinct landscape patterns: (1) encroachment into peri-urban areas resulting in sprawling cities and (2) encroachment into large expanses of greenery within the city, resulting in remnant small ordinary green spaces (Tratalos et al. 2007; Ricketts and Imnoff 2003; Kinzig and Grove 2000). Both patterns have increasingly disconnected people from the nature that supports them (Andersson 2006). To gain much-needed public involvement from multi-stakeholders for ecosystem preservation, the places where people live and work need to be designed so as to offer opportunities for meaningful interactions with the natural world (Andersson 2006; Miller 2005).

Negotiations of green spaces for development have led to sparse fragmented habitats, affecting biodiversity within the city. Such changes have allowed for rapid species turnover, extinction, reduction in specialists, and increase of generalists (Sodhi and Ehrlich 2010). For example, sparrows (*Passer domesticus*) that used to nest on house rooftops have now disappeared from the South Indian city Bangalore and are found only in the peri-urban areas (Dandapat et al. 2010). Similarly, increase in high-rise buildings has increased the number of Blue Rock Pigeons (*Columba livia*) because these buildings provide adequate nesting sites for them (Joshua and Ali 2011). In an ever-challenging task to conserve biodiversity within the city, there is a need to create and conserve green spaces which can be achieved by reconciling people's preferences with biodiversity requirements (UN-HABITAT 2010). Also, traditional theories of conservation biology focus only on large green spaces which cannot be applied any longer within cities; we need newer strategies and approaches. Reconciliation ecology which works within human-dominated ecosystem as defined by Rosenzweig is "the science of inventing, establishing and maintaining new habitats to conserve species diversity in places where people live, work or play" (Rosenzweig 2003b). There are a growing number of examples demonstrating that reconciling habitats within human-dominated landscapes has worked. For example, the US National Wildlife Federation has sponsored a campaign called "Backyard Wildlife Habitat," which encourages people to bring nature to their homes, which could vary in area from a few hectares to a single balcony. They have created and modified human habitats to provide the needs of some wildlife (Tufts and Loewer 1995) and this has even worked for endangered species such as the Eastern Blue Bird (*Sialis sialis*). It is important to focus on ordinary green spaces and change principles from the obsession of conserving rare and endangered species within urban habitats.

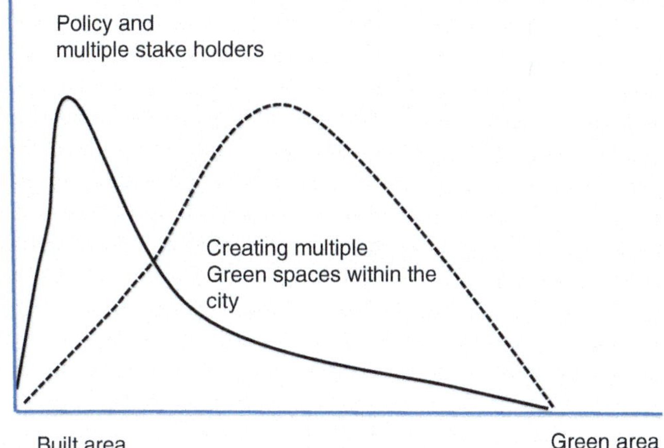

Fig. 5.1 Reconciliation ecology accomplishes biodiversity conservation and indirectly enhances the ecosystem services that small green spaces provide. (Adapted from Rosenzweig 2003a, b)

Through several such successful attempts within human-dominated landscapes, reconciliation ecology gives us hope that we can conserve and sustain habitats without necessitating a tradeoff between biodiversity and human use (Kenward et al. 2011). For example, within residential neighborhoods, creating home gardens, terrace gardens, neighborhood parks (NPs), and planting avenue trees by involving multiple stakeholders could increase the green areas as well as help conserve biodiversity (Fig. 5.1). Appropriate policies implemented effectively could also help increase the services that small green spaces provide to the community, thereby balancing green vs built-up areas within neighborhoods (Fig. 5.1). These habitats not only provide opportunities for human–nature links on a daily basis, but also deliver vital environmental services that contribute to a healthy and satisfying living environment, essential for human well-being (Millennium Ecosystem Assessment 2005).

Even if humanity is increasingly urban, we are still as dependent on the services that urban green spaces provide as before. This increasing urbanization has modified the ecology of landscapes by: changing habitats and leading to habitat fragmentation and creation of novel habitat types (Niemela 1999; Wood and Pullin 2000); altering resource flows including reduction in net productivity and increasing temperature and degradation of air and water quality (Henry and Dicks 1987; Rebele 1994; Donovan et al. 2005); shifting disturbance regimes, with many habitats experiencing frequent disruptions to development (Tratalos et al. 2007); and changing species composition and diversity (McKinney 2002). In fact, with all these escalating changes, and with increasing awareness, citizens have a growing expectation from these small green islands in terms of the range of ecosystem services (ESs) they provide such as: supporting (nutrient cycling, soil formation),

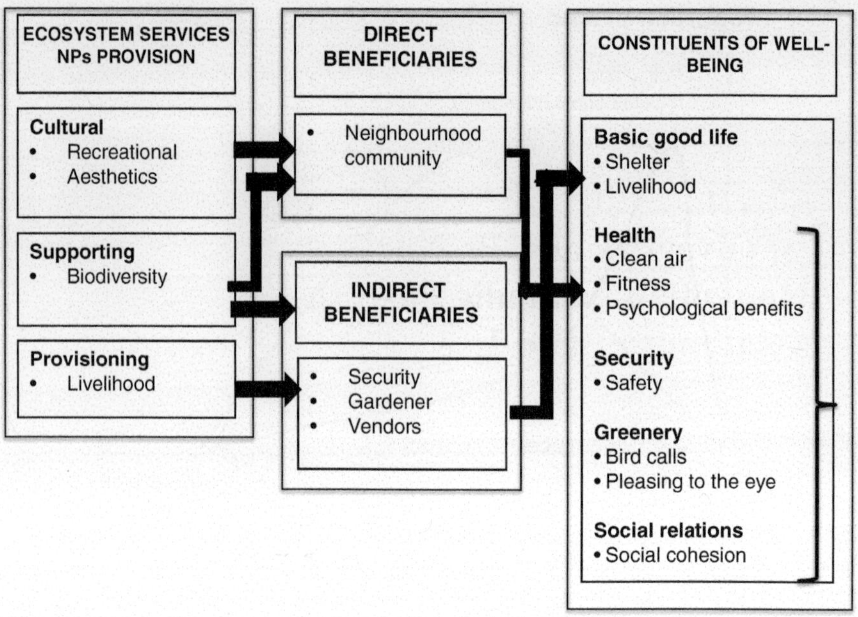

Fig. 5.2 Relationships between ecosystem services and human well-being. (Adapted from Millennium ecosystem Assessment 2005)

regulating (climate, hydrology), provisioning (food, water), biodiversity, cultural (recreation, enhancement of property value), and many others (Millennium Ecosystem Assessment 2005). Studies have shown that the ESs provided by green spaces are directly related to human well-being and sustainability of the city (Bolund and Hunhammar 1999; Daily 1997). As the Millennium Ecosystem Assessment (MEA) is about improving human well-being, from the information we have gathered the community also identifies biodiversity as a service; we therefore use a modified MEA framework throughout this study (Fig. 5.2). The community perceives biodiversity to provide it with benefits such as fruit and flowers through pollination. Thus, we choose to include biodiversity under supporting services (Fig. 5.2).

In order to increase livability, several countries are developing newer greening concepts such as vertical and roof top gardening and implementing them to provide ESs to the population and also act as corridors specifically for the mobile taxa to move from one patch to another (Getter and Rowe 2006). With the increase in people's dependence on urban green spaces for recreation/esthetics and other ESs, there is an urgent need to gear up the functioning of these spaces through conservation and appropriate management practices (Devy et al. 2009).

To achieve biodiversity conservation and to enhance ESs, multiple stakeholder involvement at various scales and policies that can be adapted with the changing environment, are needed. Urban systems have long been considering the social and the ecological systems as two different elements within a common system. Both these systems in fact are interwoven and need to be perceived as a socio-ecological system which can help us understand the resilience of such a system and strategize towards sustainable development of urban green spaces (Tarraga and Miguel 2006). For the last few years, developing countries have been attracting people from all over the world. This has resulted in exchange of cultures, leading to changes in lifestyle, attitudes, perception of people, and, to a large extent, has also influenced our urban green spaces by bringing in trendy looks (Swamy and Devy 2012). For example, people in India are switching over from traditional home gardens to manicured lawns and turf which is the natural vegetation in temperate countries. Even offices and institutional campuses are experiencing the same trend. Recently developed neighborhood parks (NPs) in Bangalore are no longer wooded stands and these changes have resulted in a cascading effect not only on the biodiversity sheltered by the traditional green spaces, but also on the processes and functions of the urban ecosystem that promote human health and well-being (Millennium Ecosystem Assessment 2005).

To conserve and maintain a resilient green space with minimal biodiversity components providing optimal ESs, one has to develop innovative partnerships, collaboration, and stewardship, which the socio-ecological system framework demands (Stringer et al. 2006). Adaptive co-management, which is "learning while doing," could be a useful model to adopt as it offers opportunities to examine the potential of collaboration between partners which integrates ecology and society (Hahn et al. 2006). Adaptive co-management also focuses on creating functional feedback loops between social and ecological systems, and has been a useful tool in tracking sustainable paths and building social-ecological resilience (Berkes and Folke 1998; Gunderson and Holling 2002). It relies on seizing a window of opportunity and linking diverse set of actors operating at different levels, often in networks from local users to municipalities to regional and national organizations (Ernstson et al. 2010). For example, residential neighborhoods comprise diverse green spaces, from NPs, avenue trees, and institution campuses to home gardens. Linking them at the neighborhood scale would require municipality, local academic institutions, and home garden owners to work together to help develop a green network at the neighborhood scale (Fig. 5.3). Perhaps modelling a socio-ecological system and projecting various paths under various scenarios based on current management will help all players visualize the future of some existing systems in cities.

In recent years, conservation and sustainability of natural resources have also been highlighted in the urban context (Newman 1999). Unlike pristine ecosystems, where unsustainable extraction has led to deterioration of resources (Ostrom 1990), in the urban context it mainly pertains to reduction in ecosystem services (ESs) provided by green spaces because of developmental activities. Following inefficient management and lack of monitoring green spaces for improvement or conversion to alternative use by the governing body responsible, urban ecologists have stressed and demanded an integrated approach, arranging for multiple stakeholders,

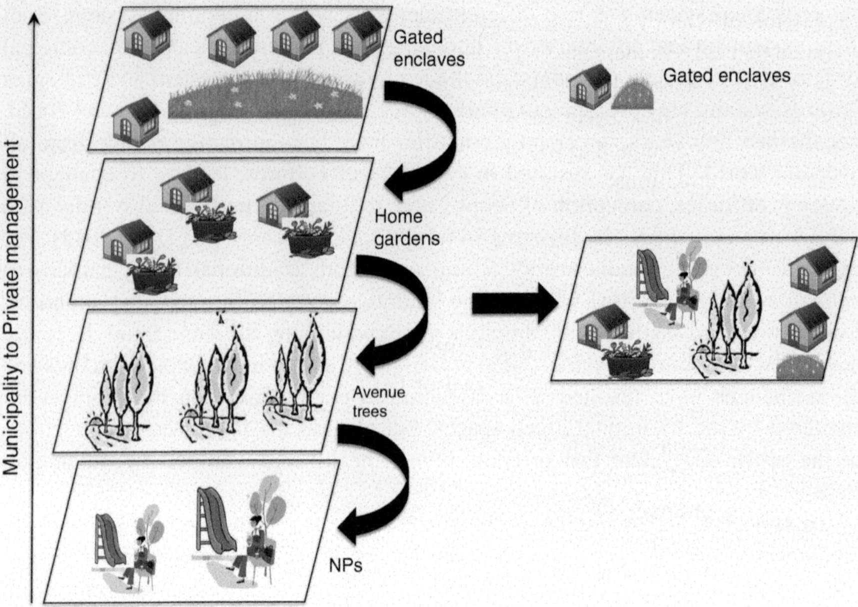

Fig. 5.3 Green network within residential neighborhoods in the presence of neighborhood parks

including citizen groups, to prevent further deterioration of urban ecological systems (Grimm et al. 2000; Pickett et al. 2001).

Bangalore city, once known as the "Garden City" of India, has seen a great level of erosion in green spaces, so much so that it has lost its earlier identity and is now looked on more as the "Silicon Valley" of India (Nagendra and Gopal 2010). Like other Asian cities, Bangalore has taken a path that faces a dilemma between conservation of green spaces and development. Although policies on urban green space exist, they are seldom enforced by the Bruhat Bengaluru Mahanagara Palike (BBMP), horticulture departments (urban governing bodies), or the citizens, and are more often violated. Policies, which lack flexibility, adaptive resilience and multi-institutional involvement, have been identified as causal factors for inefficient green space management in the urban context (Olsson et al. 2007; Olsson et al. 2004). Most of the policies related to green spaces that exist today for Bangalore were formulated several decades ago (Ravindran 2007). As policy makers did not foresee the complexities, the policies drafted remain largely outdated, as they are unable to tackle issues that are prevalent today. Apart from being outdated, these policies do not integrate feedback from stakeholders and other basic processes that regulate the dynamics of green spaces within the system in order to function better (Tarraga and Miguel 2006). Here we focus on neighborhood parks (NPs) because large green spaces within the city receive exceptional patronage from the citizen groups, naturalists, and several other stakeholders. In contrast, although NPs are vital green spaces within neighborhoods, they are completely neglected and lack public empathy; hence they are constantly under threat of alternative uses such as institutional playgrounds and civic amenity centers (Swamy and Devy 2010). Rosenzweig (2003a) suggests that it is essential to reconcile nature even in places where people live, work, and play. Thus, this chapter delves into the premise of transforming NPs, which are emerging as necessary "urban commons" at a neighborhood scale, to support at least "ordinary nature". We also provide a framework based on some of our findings to make NPs multifunctional.

5.2 Key Findings

Ecosystem services—Integrating the community's requirements such as esthetic and recreational services, mixed landscape types that people prefer to open or compact parks, increasing density of NPs in the neighborhood and enhancing biodiversity beginning with people's most-liked taxa (Swamy 2013; Fig. 5.4). A biodiversity fondness survey was conducted, and the Likert-scaling method was employed to assess people's tolerance levels towards nine commonly encountered taxa in NPs (Meyers et al. 2005). A total of 425 questionnaire surveys were conducted amongst park users to identify their fondness across 9 taxa that were

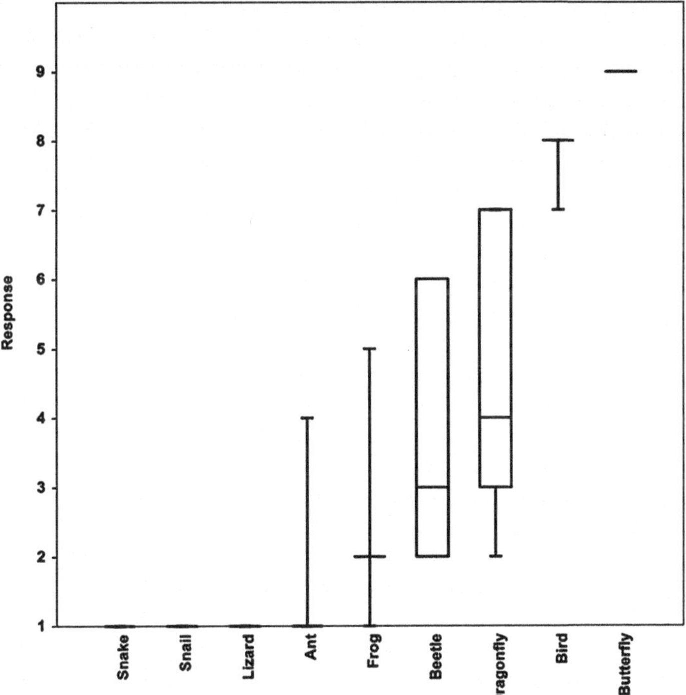

Fig. 5.4 Fondness survey across nine commonly found taxa within neighborhood parks using the Likert scale where 1 represents complete intolerance towards the taxa and 9 extreme fondness

commonly found in NPs. A nine-level scaling method (Dawes 2008) was used, where nine denotes fondness towards the taxa and one indicates complete intolerance. The Likert-scaling helped identify the two taxa that people were fond of—birds and butterflies—which were then sampled systematically in all 37 NPs, using the point count method. The results showed that users were fond of butterflies followed by birds. Some taxa such as lizards, snails, and snakes belonged to the intolerant scale (Fig. 5.4) and were disliked by all (100 %) respondents.

Biodiversity support services—Based on the density and sizes of parks, four distinct groups were recognized—high-density NPs along with the presence/absence of large green spaces (HNP + LP; HNP-LP) and low-density NPs in the presence/absence of large green spaces (LNP + LP; LNP-LP). The species accumulation curve across three classes of parks embedded within a 2-km buffer was compared for birds, butterflies, and insects, and showed varying NP densities within a 2-km buffer. High density packing of NPs in the absence of large parks (HNP-LP) and low density of NPs in the presence of large parks were high in species richness (LNP + LP; Fig. 5.5). LNP-LP showed a lower level of species accumulation; there were very few locations in the city with HNP + LP.

Creating high-density NPs and/linking sparse NPs to the large green space in the vicinity could help enhance biodiversity support service within NPs and their neighborhoods. Linking several small green spaces can also increase local species richness within residential neighborhoods (Fig. 5.5).

Management—Linking multiple stakeholders could help distribute responsibilities and increases efficiency and knowledge to help create a resilient system, such as involving the Residential Welfare Association (RWA) in managing the park in collaboration with the municipality along with the participation of ecological organizations. These were classified as (Fig. 5.6) co-managed NPs (henceforth CoM NPs)—the tenure of all the parks within the BBMP boundary is managed and owned by the BBMP horticulture department. Through the "adoption policy," as stated by the BBMP horticulture department, interested individuals within a few areas have formed an RWA, a statutory body, which has collaborated with the BBMP horticulture department in managing the parks within their neighborhood. This was compared with city-managed NPs (henceforth CiM NPs). The BBMP horticulture department manages and maintains the NP through its employees who are gardeners and landscape contractors. Social Network Analysis (henceforth SNA) is a useful tool to study effectively the governance organizations, comprising individual actors who are linked together through various relationships (Scholz and Wang 2006; Bodin et al. 2006; Crona and Bodin 2009; Ernstson et al. 2010). A survey started with the president of the RWA and followed a snowball effect until the complete network of actors was covered.

Using SNA, this study delineated the management structures responsible for NPs and identified gaps and means to strengthen the networks from the current state to the near-ideal state so as to provide enhanced services through better green space governance (Fig. 5.6). The numbers of actors (individuals) within the CoM NPs seem to differ across the replicates, hence allowing for structural variations and adaptive flexibility unlike CiM NPs, which are uniform The CiM replicates show

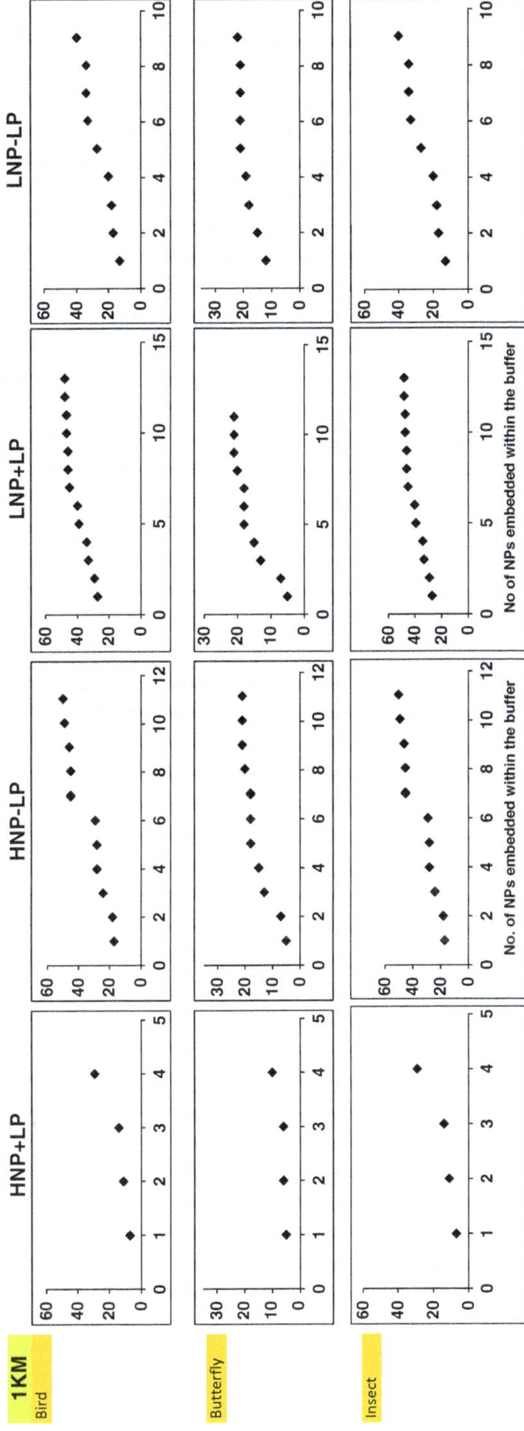

Fig. 5.5 Species accumulation with increasing NPs in buffer of 1 km with various combinations

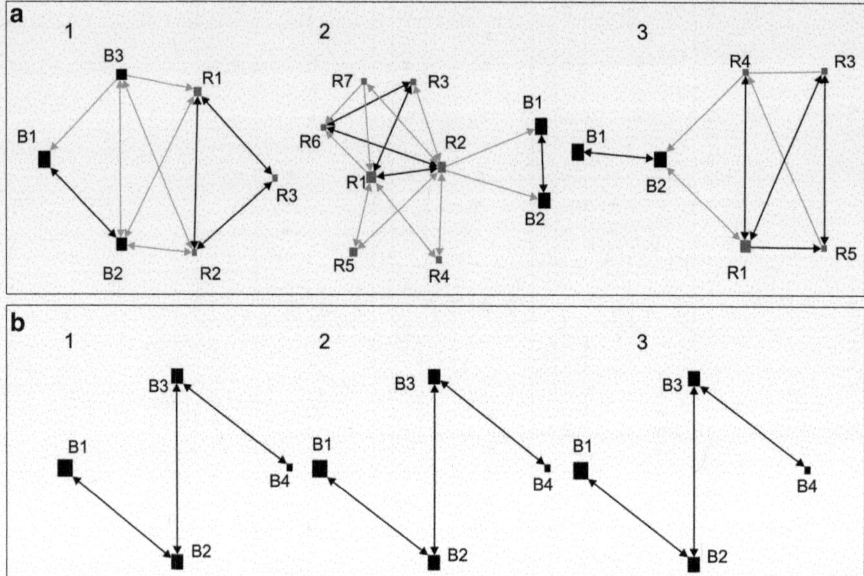

Fig. 5.6 Involvement of multiple stakeholders across replicates demonstrating strength (represented by *thickness*) and flexibility. R (in *gray*) represents actors within the RWA and B (in *black*) represents actors who belong to the BBMP horticulture department. *Thicker lines* represent higher levels of interactions between actors. *Symbols* represent an actor who is a single person, size of the symbol represents his/her position in network

that the links between all actors are one-way. The absence of two-way links clearly demonstrates the strict hierarchical system within the BBMP horticulture department (Fig. 5.7) in governance and management; hence, showing a linear top-down relationship within CiM NPs and among the BBMP actors belonging to CoM NPs (Fig. 5.7).

Governance and policy—A bottom-up approach, with a feedback mechanism in place could help develop a better green space management for NPs. The absence of two-way links among the municipality employees clearly demonstrates the strict hierarchical system within the BBMP horticulture department in the governance and management thereof; hence, showing a linear top-down relationship (Fig. 5.7). Involving the Resident Welfare Association (RWA) could help in complex linkages (Fig. 5.6), rendering the management more flexible and strong both in terms of knowledge and interaction to make efficient decisions that feed back into the system.

The existing policies are reviewed and a conceptual model for pocket green space is proposed at neighborhood scale based on the findings on ESs provided by NPs and appreciated by the citizens in Bangalore. The results showed that although NPs provide intangible services that are essential for human well-being (Bolund and Hunhammar 1999), they constantly face threats of being replaced by development activities. Lack of stewardship to safeguard these pocket green spaces has led to easy conversion to alternative uses. General apathy towards NPs could stem from the fact that the "vital services" provided by them have not been highlighted

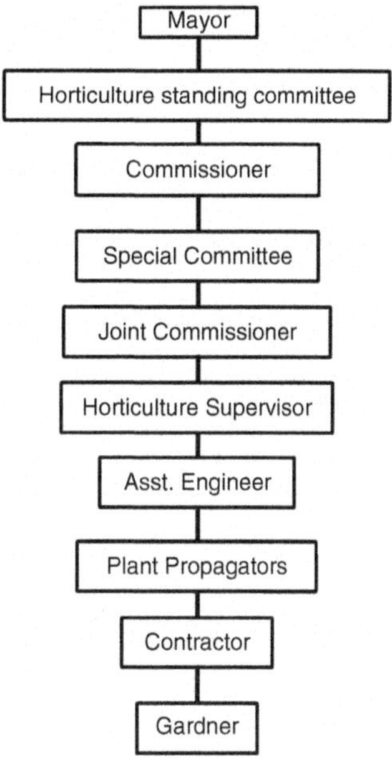

Fig. 5.7 Hierarchical chart of management personnel within the BBMP horticulture department

vis-à-vis large heritage parks (henceforth HPs), which by their sheer size and long duration of presence in the city have received larger patronage by citizens and conservationists. Although, HPs provide ESs similar to those of NPs but at a higher magnitude, they have also attracted the "nature enthusiasts" because of the higher biodiversity they support. All articles from two popular and well circulated dailies, *The Hindu*, *Deccan Herald* and *Times of India*, regarding threats to green spaces were collected between 2007 June and 2012 May, and these were used to assess the stakeholders' contributions through participation in protests towards diverse green spaces within Bangalore city.

Results obtained from a total of 122 newspaper articles clearly suggest that there is an upsurge of citizenry involvement whenever there is a threat to large green spaces and avenue trees, but threats towards NPs go unnoticed and do not gain so much publicity (Fig. 5.8). The newspaper article survey over a period of 6 years has shown that only two NPs were put to alternative use. In contrast to this, interviews with the park target groups revealed that, in several areas, plots allotted for NPs have been converted into civic amity centers or playgrounds. Bangalore saw the major infrastructure expansion from 2009 and 2011 which saw rampant expansion of roads and clearing of large parks for mass transit systems. This brought citizens and environmental groups to protest in the street. This served as feedback to municipality and has brought some respite to the clearing of avenue trees and large green spaces. There is also general apathy among the citizens towards neighborhood parks, although they are used on a daily basis.

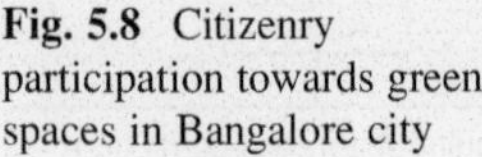

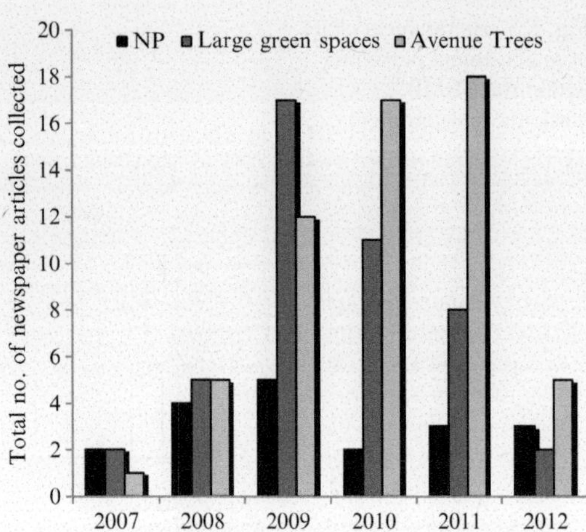

Fig. 5.8 Citizenry participation towards green spaces in Bangalore city

5.3 Conceptual Model

A conceptual framework is proposed by us for sustainability of NPs based on findings (Fig. 5.9).

Integrating and gradually converting the dominant open type (trees only along the boundary of the park) NPs to mixed landscape type (trees along the boundary wall and scattered in the center of the park), with taxa-specific habitat features such as canopy cover, shrub abundance, and herb proportion within NPs and in the surrounding landscape, could enhance local species richness within NPs and at the neighborhood scale. Increasing neighborhoods with higher densities of NPs than are prevalent now could not only facilitate diverse recreational services for the community but also help achieve conserving biodiversity in the neighborhood. Our analysis on determinants of biodiversity support services showed that a high density of NPs can be effective even in the absence of large parks, while a low density of NPs in the presence of large parks can enhance local species richness (Swamy 2013). Linking other neighborhood green spaces such as home gardens, avenue trees and NPs could also effectively help enhance biodiversity at the neighborhood scale (Fig. 5.3).

A bottom-up approach which incorporates not just popular services such as esthetic and recreation, but also biodiversity services, by involving multiple stakeholders, allowing for both direct and indirect feedbacks into the system; would help develop better green space policies.

This can be a daunting task if the governance organizations responsible for neighborhood green spaces are absent or are not efficient in managing these critical spaces. Thus, the governance organizations responsible for green spaces within neighborhoods could play an important role in enhancing and building green networks. Knowledge on NPs can provide enhanced services, but the dominant governance structure, which is driven by the municipality, continues to manage them without knowledge of the biodiversity perspective and people's requirements. Discontentment amongst the citizen group with the inability of the municipality to

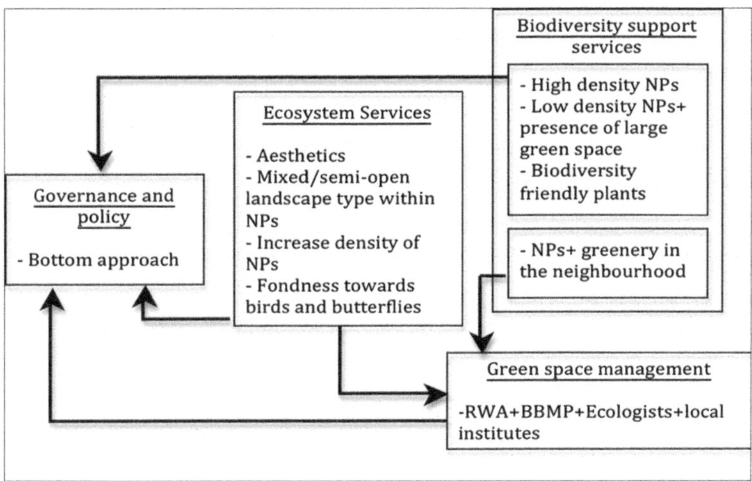

Fig. 5.9 Conceptual framework by linking ecosystem services around neighborhood parks

manage NPs, has led them to form a group through a Residential Welfare Association (RWA). Such stewardship has also helped prevent conversion of a few NPs into alternative use (Anonymous 2001). RWAs have collaborated with the municipality to improve the management of NPs through co-management (Fig. 5.6). This newly emerging governance structure has, to an extent, allowed citizens to incorporate facilities within NPs according to their requirements. Lack of ecological knowledge amongst the RWA members and the municipality does not allow for enhancing the ESs of NPs. Thus, to strengthen the co-management around NPs requires involvement of ecologists and conservation biologists for sharing knowledge, building the capacity of the neighborhood community for growing biodiversity friendly plants within homes, installing accessories such as nest boxes, and, finally, involving local academic institutions such as schools and colleges in long-term monitoring programs with the municipality. This will help multiple stakeholders to come together onto a common platform to scale up NPs to provide enhanced esthetic, recreational and biodiversity support services.

References

Andersson E (2006) Urban landscapes and sustainable cities. Ecol Soc 11:34. Available via DIALOG. http://www.ecologyandsociety.org/vol11/iss1/art34/

Anonymous (2001) Sadashivanagar association. A residential association report

Berkes F, Folke C (1998) Linking social and ecological systems. Cambridge University Press, Cambridge

Bodin O, Crona B, Ernstson H (2006) Social networks in natural resource management: what is there to learn from a structural perspective? Ecol Soc 11:r2

Bolund P, Hunhammar S (1999) Ecosystem services in urban areas. Ecol Econ 29:293–301
Crona B, Bodin O (2009) The role of social networks in natural resource governance: what relational patterns make a difference? Glob Environ Chang 19:366–374
Daily GC (1997) Introduction: what are ecosystem services. In: Daily GC (ed) Nature's services: societal dependence on natural ecosystems. Island Press, Washington, DC, pp 1–10
Dandapat A, Banerjee D, Chakraborty D (2010) The case of the disappearing house sparrow (*Passer domesticus indicus*). Vet World 3:97–100
Dawes J (2008) Do data characteristics change according to the number of scale points used? An experiment using 5-point, 7-point and 10-point scales. Int J Market Res 50:61–77
Devy MS, Swamy S, Aravind NA (2009) Reshaping urban green spaces. Econ Polit Wkly 44:25–27
Donovan RG, Stewart HE, Owen SM, MacKenzie AR, Hewitt CN (2005) Development and application of an urban air quality score for photochemical pollution episodes using the Birmingham, United Kingdom, area as a case study. Environ Sci Technol 39:6730–6738
Ernstson H, Barthel S, Andersson E, Borgström ST (2010) Scale-crossing brokers and network governance of urban ecosystem services: the case of Stockholm. Ecol Soc 15:28. Available via DIALOG. http://www.ecologyandsociety.org/vol15/iss4/art28/.
Folke C, Jansson A, Larsson J, Costanza R (1997) Ecosystem appropriation by cities. Ambio 26:167–172
Getter KL, Rowe DB (2006) The role of green roofs in sustainable development. Hortic Sci 41:1276–1285
Grimm NB, Grove JM, Pickett STA, Redman CL (2000) Integrated approaches to long-term studies of urban ecological systems. Bioscience 50:571–584
Gunderson LH, Holling CS (2002) Panarchy: understanding transformations in human and natural systems. Island Press, Washington, DC
Hahn TP, Olsson P, Folke C, Johansson K (2006) Trust-building, knowledge generation and organizational innovations: the role of a bridging organization for adaptive comanagement of a wetland landscape around Kristianstad, Sweden. Hum Ecol 34:573–592
Henry JA, Dicks SM (1987) Association of urban temperatures with land-use and surface materials. Landsc Urban Plan 14:21–29
Joshua G, Ali Z (2011) Avian diversity with the varying urban congestion of Lahore. J Anim Plant Sci 21:421–482
Kenward RE, Whittingham MJ, Arampatzis S, Manos BD, Hahn T, Terry A, Simoncini R, Alcom J, Bastian O, Donlan M, Elowe K, Franzen F, Karacsonyi Z, Larsson M, Manou D, Navodaru I, Papadopoulou O, Papathanasiou J, von Raggamby A, Sharp RJA, Soderqvist T, Vavrova L, Aebischer NJ, Leader-Williams N, Rutz C (2011) Identifying governance strategies that effectively support ecosystem services, resource sustainability and biodiversity. Proc Indian Acad Sci 180:5308–5312
Kinzig AP, Grove JM (2000) Urban-suburban ecology. In: Levin S (ed) The encyclopedia of biodiversity. Academic, San Diego, pp 733–746
McKinney ML (2002) Urbanization, biodiversity and conservation. Biosci 52:883–890
Meyers LS, Guarino A, Gamst G (2005) Applied multivariate research: design and interpretation, 2nd edn. Sage, Thousand Oaks
Millennium Ecosystem Assessment (2005) Ecosystems and human well-being. Island Press, Washington, DC
Miller JR (2005) Biodiversity conservation and the extinction of experience. Trends Ecol Evol 20:430–434
Nagendra H, Gopal D (2010) Tree diversity, distribution, history and change in urban parks: studies in Bangalore, India. Urban Ecosyst 14:211–223
Newman PWG (1999) Sustainability and cities: extending the metabolism model. Landsc Urban Plan 44:219–226
Niemela J (1999) Ecology and urban planning. Biodivers Conserv 8:119–131

Olsson P, Folke C, Hahn T (2004) Socio-ecological transformation in ecosystem management: the development of adaptive co-management of a wetland in Southern Sweden. Ecol Soc 9:2

Olsson P, Folke C, Galaz V, Hahn T, Schultz L (2007) Enhancing the fit through adaptive co-management: creating and maintaining bridging functions for matching scales in the Kristianstads Vattenrike Biosphere Reserve, Sweden. Ecol Soc 12:38

Ostrom E (1990) Governing the commons: the evolution of institutions for collective action. Cambridge University Press, New York

Pickett STA, Cadenasso ML, Grove JM, Nilon CH, Pouyat RV, Zipperer WC, Costanza R (2001) Urban ecological systems: linking terrestrial, ecological, physical and socioeconomic components of metropolitan areas. Annu Rev Ecol Syst 32:127–157

Ravindran DS (2007) Wither open spaces? A politico-economic analysis of open space provisioning in Bangalore, India. Unpublished PGPPM Thesis, Indian Institute of Management, Bangalore

Rebele F (1994) Urban ecology and special features of urban ecosystems. Global Ecol Biogeogr Lett 4:173–187

Rees WE (1997) Urban ecosystems: the human dimension. Urban Ecosyst 1:63–75

Rees WE (2003) Understanding urban ecosystems: an ecological economics perspective. In: Berkowitz AR, Nilon CH, Hollweg KS (eds) Understanding urban ecosystems. A new frontier for science and education. Springer-Verlag, New York, pp 115–136

Ricketts T, Imnoff M (2003) Biodiversity, urban areas and agriculture: locating priority ecoregions for conservation. Conserv Ecol 8:1

Rosenzweig ML (2003a) Reconciliation ecology and the future of species diversity. Oryx 37:194–205

Rosenzweig ML (2003b) Win-win ecology: how the Earth's species can survive in the midst of human enterprise. Oxford University Press, New York

Scholz JT, Wang CL (2006) Cooptation or transformation? Local policy networks and federal regulatory enforcement. Am J Polit Sci 50:81–97

Sodhi NS, Ehrlich P (2010) Conservation biology for all. Oxford University Press, Oxford

Stringer LC, Dougill AJ, Fraser E, Hubacek K, Prell C, Reed MS (2006) Unpacking "participation" in the adaptive management of socio-ecological systems: a critical review. Ecol Soc 11:39

Swamy S (2013) Reshaping neighbourhood parks for biodiversity and people: a study on pocket green spaces in Bangalore, India. Ashoka Trust for Research in Ecology and the Environment (ATREE), Bangalore, PhD thesis

Swamy S, Devy S (2010) Forests, heritage green spaces, and neighborhood parks: citizen's attitude and perception towards ecosystem services in Bengaluru. Res Energy Dev 7:117–122

Swamy S, Devy S (2012) Bangalore from a "bean to boom" city: transitions of green spaces and associated cultures of a growing metropolitan in India. *Under Review*

Tarraga O, Miguel A (2006) A conceptual framework to assess sustainability in urban ecological systems. Int J Sustain Dev World Ecol 15:1–15

Tratalos J, Fuller RA, Warren PH, Davies RG, Gaston KJ (2007) Urban form, biodiversity potential and ecosystem services. Landsc Urban Plan 38:307–317

Tufts C, Loewer P (1995) Gardening for wildlife. Rodale Press, Emmaus

UN-HABITAT (2010) Planning sustainable cities: UN – habitat practices and perspectives. United Nations human settlements programs. UN-HABITAT, Kenya

Vitousek PM, Mooney HA, Lubchenco J, Melillo JM (1997) Human domination of Earth's ecosystem. Science 277:494–499

Wood BC, Pullin AS (2000) Conservation of butterflies in the urban landscape in the West Midlands. In: URGENT Annual Meeting 2000 Proceedings, Cardiff University

Part III
Adaptation to Unpredictability

Chapter 6
Quantitative Predictions for Ecological and Economic Sustainability in Mongolian Pastoral Systems

Satoshi Kato

Abstract Since the dramatic transition in the 1990s from central planning to a market-based economy in Mongolia, modern pastoralism has transformed into a more sedentary system with increasing livestock densities. Following this shift, herders were able to live in the place of their choosing; many migrated to peri-urban areas in the central part of the country, where they increased their livestock numbers to maximize profits, resulting in overgrazing and land degradation. To minimize degradation associated with such shifts in Mongolian pastoralism, an understanding of how economic and social factors can affect human activities and environmental changes via pastoralism is critical. Herein, the author developed a spatially explicit simulation model that combined the behaviors of local people with vegetation processes according to precipitation patterns in Mongolian grasslands to predict grassland degradation after 30 years. This model was parameterized with empirical data relating to Mongolian pastoralism, including rainfall patterns, changes in grassland biomass, and spatiotemporal movement patterns of nomadic peoples. The model was simulated using four scenarios that combined a mobility mode (nomadic or sedentary mode) with a specified degree of grazing pressure (high or moderate). The results suggest that mobility is a key factor for environmental and economic sustainability in Mongolian pastoral systems because, in an unpredictable environment, nomadic pastoralism was more sustainable and profitable compared to sedentary pastoralism. However, pastoral systems with excess grazing pressure were unsustainable, regardless of mobility mode. These predictions have important implications for the development of effective management strategies in the sustainability of pastoral systems.

Keywords Nomadic grazing • Scenario comparison • Sedentary grazing • Simulation • Spatially explicit agent-based model

S. Kato (✉)
Research Institute for Humanity and Nature, Kyoto, Japan

FUJIREBIO Inc., Tokyo, Japan
e-mail: kato.satoshi.0@gmail.com

S. Sakai and C. Umetsu (eds.), *Social–Ecological Systems in Transition*, Global Environmental Studies, DOI 10.1007/978-4-431-54910-9_6,

6.1 Introduction

It is commonly recognized that ecosystems across the planet have been degraded by anthropogenic activities, and many are now in a critical condition. Suitable solutions to mitigate this global crisis will require a comprehensive understanding of human socioeconomic and ecological systems, as well as the feedbacks between these systems (Ostrom 1990; Ostrom and Cox 2010; Yamamura 2013). Although many studies have focused on how human activities impact ecosystems from an ecological context, few have attempted to predict ecosystem changes related to widespread societal transformations and subsequent deviations from previously established interactions between human activities and natural resources.

In Mongolia, since the dramatic transition from central planning to a market-based economy in the 1990s, modern pastoralism has been transformed into a more sedentary system with increasing livestock (Bruun and Odgaard 1996; Kevin 2003; Pomfret 2000). Following this shift, herders were able to live in a place of their choosing, and many migrated to peri-urban areas in the central part of the country (Batjargal 1997). There they increased their livestock numbers and densities in an effort to maximize their profits. It has been shown that overgrazing due to high livestock density can be an important factor of pasture land degradation such as desertification (Badarch and Ochirbat 2002; Lise et al. 2006; MoFALI 2010). In addition, a new Pasture Law, which would allow the private ownership of pastureland and constrict nomadism, is currently in debate within the Mongolian Parliament and has led to controversy (Kamimura 2013). The adoption of sedentary pastoralism has important implications for continuing land degradation and is of growing concern in this landscape (Fernandez-Gimenez 2001).

Understanding how economic and social factors can affect human activities and environmental changes via pastoralism is critical for minimizing the degradation associated with such shifts in Mongolian pastoralism. In this study, the author developed a spatially explicit simulation model that combined behaviors of nomadic people with vegetation processes according to precipitation patterns to predict pasture productivity and grassland degradation over 30 years in Mongolia. Sustainable pastoralism as an industry requires not only a healthy grassland ecosystem, but also economic viability. The objectives of this study were to analyze how mobility mode (nomadic or sedentary) and grazing pressure (high or moderate) impacted land degradation and agricultural sustainability in this pastoral system.

6.2 Materials and Methods

The simulation model used in this study consisted of two components, describing (1) vegetation processes and (2) pastoral events. The vegetation processes involved the growth of aboveground biomass after precipitation, grazing by livestock, and changes in soil condition. The pastoral events consisted of herder settlement time

Table 6.1 Parameters and their sources used in the simulation model

Category	Data definition	Method/source
Precipitation	Long-term frequency (mm/15 days)	National Statistical Office (Ulaanbaatar)
	Spatial variation	Experimental measurement
Vegetation	Response in initial biomass to - rainfall (kg/mm)	Experimental measurement
	Response in growth rate to rainfall (kg/mm/15 days)	Experimental measurement
Pasture	Range of dairy pasture (km)	GPS measurement
	Distribution of grazing pressure on dairy pasture	GPS measurement
	Moving distance from household during grazing (km)	GPS measurement
	Number of households (households)	National Statistical Office (Ulaanbaatar)
	Number of livestock heads (as biomass; 50 kg/head[a])	National Statistical Office (Ulaanbaatar)
	Total area of prefecture (km^2)	National Statistical Office (Ulaanbaatar)
	Loss of goats by natural mortality (heads/region)	National Statistical Office (Ulaanbaatar)
	Total livestock for consumption (heads/region)	National Statistical Office (Ulaanbaatar)
	Grazing rate for single head of livestock (kg/15 days/head)	Experimental measurement
	Transfer efficiency for livestock (kg/kg)	Holmes and Jones (1964)

[a] Livestock calculations are based on weight: cows (300 kg/head), sheep (60 kg/head), and goats (50 kg/head)

on pasture area, daily pasture use, growth and consumption of livestock by each herder, and annual migration of herders (in some scenarios). This model was parameterized with recently collected empirical data describing precipitation patterns, changes in grassland biomass, and spatiotemporal daily movement patterns of nomadic people (Table 6.1, Figs. 6.1, 6.2, 6.3, and 6.4). We used a spatially explicit lattice model (120×120 cells for 60×60 km of pasture area) with every 15 days being a unit of one step in the pasture season to describe vegetation processes during the pasture season for each year. The pasture season was regarded as the beginning of April to the end of October.

The first pasture event in each step was calculated as the average amount of accumulated precipitation in a 20×20 km area every 15 days. In this model, the amount of precipitation over 15 days during pasture season was chosen from long term data for daily precipitation (millimeters/15 days) recorded by the National Statistical Office (Fig. 6.1). Then the amount of precipitation in a 5×5 km area within a 20×20 km area, i.e., spatial heterogeneity in 4×4 areas, were randomly chosen from a normal distribution with average precipitation as the amount of precipitation within a 20×20 km area and its standard deviation. The standard deviation was determined by the linear regression for the standard deviation of accumulated precipitation in each 5×5 km area against average precipitation,

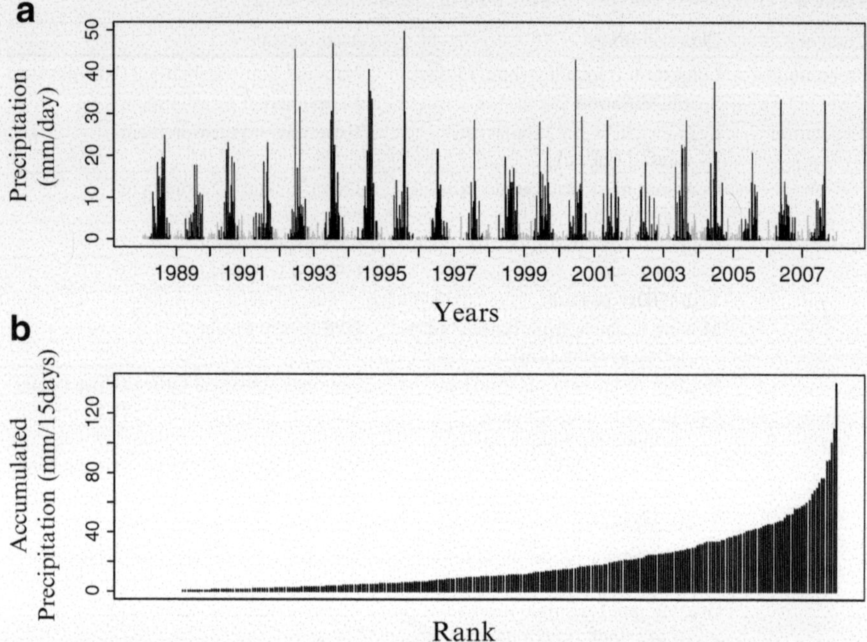

Fig. 6.1 Temporal rainfall patterns in an area of moderate grazing (Bayan-Önjüül) in Mongolia showing (**a**) daily rainfall amounts recorded by the National Statistical Office, Ulaanbaatar (April 1, 1996 to October 31, 2010), and (**b**) the frequency of accumulated rainfall amounts over continuous 15-day periods throughout the pasture season (January 1, 1996 to December 21, 2010). In (**a**), the *black* and *gray bars* show the amount of daily precipitation during the pasture and non-pasture seasons, respectively

based on the measurements of spatial variation in rainfall (Fig. 6.2). The variation of precipitation in each cell within each 5×5 km area was assumed to be homogeneous.

After determining the amount of precipitation in each cell, initial grassland biomass at the beginning of the pasture season and the growth rate during the pasture season were calculated based on both the amount of precipitation and the degree of land degradation. Initial biomass at first sprout of pasture season was determined by the total amount of precipitation and relative slope, determined by the condition of land degradation in each cell. Figure 6.3 shows data from three localities. As shown in Fig. 6.3a, in all localities there was no initial production in biomass when the accumulated precipitation amount over 15 days was <10 mm. Vegetation responded proportionally with precipitation amounts exceeding this level, but the slopes were different between localities. We regarded this difference as the degree of land degradation. Thus, the relationship between the amount of precipitation and initial biomass is described as

$$b_{k,t} = \max\left(0, B_{\max} u_{k,t} \left(p_{k,t} - 10\right)\right),$$

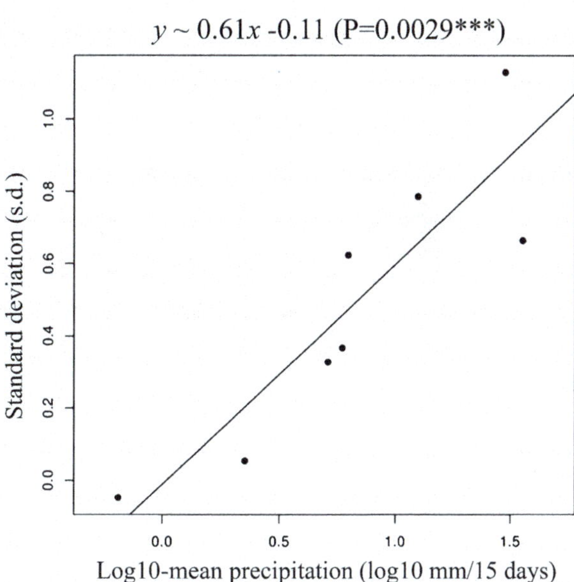

Fig. 6.2 Linear regression between standard deviations (SD), i.e., spatial variations in precipitation among 4 × 4 points of measure against average accumulated rainfall over continuous 15-day periods within a 20 × 20 km area

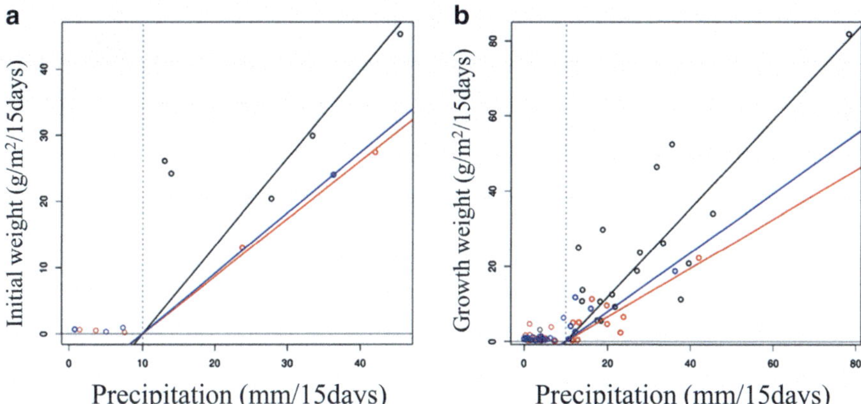

Fig. 6.3 Changes in (**a**) initial biomass and (**b**) growth rate of grasslands according to variations in accumulated rainfall amounts over continuous 15-day periods throughout Mongolia, including the cities of Gachuurt (*black*), Mandalgovi (*red*), and Dalanzadgad (*blue*). Grassland responses in Gachuurt were assumed to be the physiological maxima in this ecosystem for all model simulations

where $B_{\max}$ is the maximum regression slope of initial biomass against precipitation, $u_{k,t}$ is relative to maximal slope, and $p_{k,t}$ is the total amount of precipitation over 15 days between t and $t-1$ in a specific cell (i, j) at the t_{th} time step in the k_{th} year, respectively. Growth of aboveground biomass was started once initial biomass

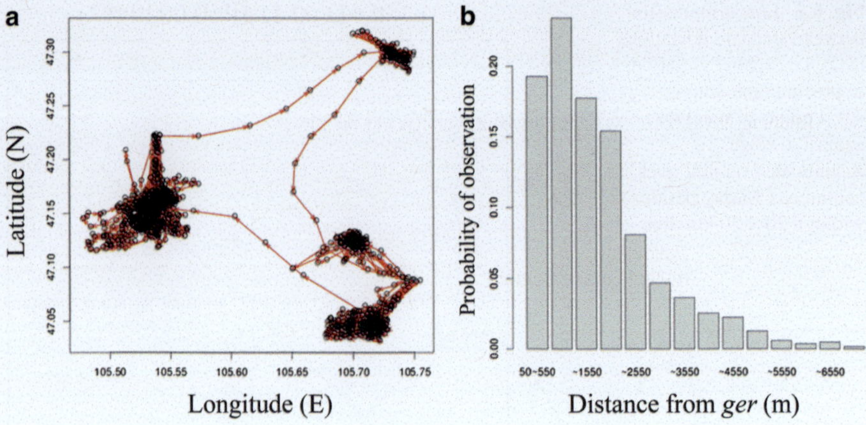

Fig. 6.4 Grazing patterns for a nomadic pasture system, including (**a**) daily movement patterns of a nomadic herder as measured through GPS (May 4, 2009 to August 23, 2009; *open circles*) and (**b**) the probability of a grazing event as a function of distance from a herder's household (*ger*; May 4, 2009 to August 23, 2009). The direction and length of each *red arrow* in (**a**) is indicative of a nomad's relative movement direction and distance, respectively, in every other hour of grazing. In (**b**), the probability was counted as the number of events within a 500-m width of a concentric circle. The distance is from the *ger* to the outer circumference

hatched, and, accordingly, $p_{k,t}$ exceeded 10 (mm/15 days). Because pastoral events were designed to begin 1 month from the start of the vegetation process in the pasture season, aboveground biomass ($b_{k,t}$) increased in response to the amount of rainfall and decreased with livestock grazing as follows:

$$b_{k,t+1}^{i,j} = \min\left(b_{\max}, g_{k,t} \cdot b_{k,t}\right)^{i,j} \text{ if } t = 1 \text{ or } 2,$$

$$b_{k,t+1}^{i,j} = \min\left(b_{\max}, g_{k,t} \cdot b_{k,t} - \sum_{x \in \text{herder}} F_x\right)^{i,j} \text{ if } t > 3,$$

Here, $g_{k,t}$ and F_x are the growth rate and consumption of aboveground grassland biomass by each herder, respectively. The details of F_x are described later in pastoral events. Similar to initial biomass, referring to data from three locations, the growth rate of aboveground biomass was also given as

$$g_{k,t} = R_{\max} u_{k,t} \left(p_{k,t} - 10\right),$$

where $R_{\max}$ is the maximal regression slope of vegetation growth against the total amount of precipitation (Fig. 6.3b). Both maximum slopes, $B_{\max}$ and $R_{\max}$, were estimated through observations at Gatuurt city, 400 km south of Ulaanbaatar (UB), which was assumedly characterized by optimal soil conditions. To evaluate environmental sustainability, the degree of degradation of pasture area (DDPA) is defined as

$$\mathrm{DDPA} = 1 - u.$$

It was assumed that DDPA increased with total primary production at a fixed conversion efficiency k and decreased at a fixed loss rate v over time. It was also assumed that DDPA increased with the removal of aboveground biomass at the end of each pasture season. Using these assumptions, the relative coefficient to maximal slope at step $t+1$ in k_{th} year ($u_{k,t+1}$) and initial states at the beginning of the next year ($u_{k+1,0}$) were calculated from the previous state ($u_{k,t}$ and $u_{k,\,\mathrm{last}}$, respectively) as the following equations:

$$u_{k,t+1} = (1 - v)u_{k,t} + kg_{k,t}b_{k,t}$$
$$u_{k+1,0} = (1 - v)u_{k,\mathrm{last}} + kb_{k,\mathrm{last}}$$

All simulations were initiated at DDPA = 0.2 for each cell. It was assumed that 80 % of $R_{\max}$ would be kept under that of the current grazing pressure in Bayan-Önjüül. On the basis of 100 replications of preliminary simulations without changes in livestock biomass, the degradation rate ($v = 0.02$), and conversion efficiency ($k = 0.04$) were determined. The decrease in aboveground biomass was calculated based on the number of livestock that each herder owned and by the spatial distribution of grazing pressure around the house of the herder, called a *ger*, at a spatial resolution of 500 × 500 m.

Pastoral events were described considering a *ger* as a constitutional unit of the individual-based model. In the model, each herder had a specific *ger* location and a set number of livestock. The number of livestock was calculated by goat weight (50 kg/head) from total biomass. Total biomass of livestock was increased with the amount of feeding during the pasture season, which was transferred with conversion efficiency c to newly-born livestock at the start of the next season. After the pasture season, the number of livestock decreased with natural mortality ($m = 0.03$), sales, and self-consumption ($h = 0.2$) each year. Changes in total livestock biomass (kg) owned by a specific herder (w_k) were described according to the following equation:

$$w_{k+1} = (1 - h)\left\{(1 - m-)\left(w_k + c\sum_t F_t\right)\right\}.$$

Herder mobility concerning land tenure on different grazing pressures was compared; one case is that herders continue to relocate to pasture in a shared area, and the other is that herders shift to sedentary pasture because of land tenure conflicts brought about by the new Pasture Law. Simulations were conducted in four scenarios with varied mobility mode and grazing pressure. The mobility mode could be "nomadic pastoralism" or "sedentary pastoralism." Under the nomadic system, herders were randomly placed in pasture areas at the beginning of the pasture season. After every 15 days of occupation with daily pasture, each herder could continue to stay at his current location or move to a new location. Every 15 days in pasture season, each herder tried to find a more profitable location than the current one based on the amount of vegetation. In this simulation, we assumed that a herder compared his current location with three other sites within 21 km of his *ger*. The assumption for searching distance for candidate location was based on maximal migration distance from GPS logging (Fig. 6.4). The distribution of grazing pressure in 15 days per herder was not uniform in the nomadic system. It was assumed that the closer it is to the center of the circle (i.e., setting location of *ger*) the more grazing pressure increases. This weight was calculated from real data for GPS logging corresponding to the hourly movements of real herders (Nachinshonhor and Jargalsaikhan 2013; Fig. 6.4b). On the other hand, under the sedentary system every herder was assumed to inhibit annual migration and to be distributed at fixed locations throughout the pasture area where they exclusively grazed their animals during the full pasture season. We assumed that the same herder used the same location through 30 years, which is contrary to the nomadic system. In addition, we assumed that the grazing pressure was uniform throughout the pasture occupied by each herder.

Grazing pressure was based on the statistical accounts in Mongolia (National Statistical Office, Ulaanbaatar), where high grazing pressure occurred in UB, with 1,600 households grazing 213.1 goats per household. Moderate grazing pressure occurred in Bayan-Önjüül (BO), with 529 households grazing 397.3 goats per household.

Under each scenario, we calculated the number of livestock held by each herder, as well as DDPA every year for 30 years. The number of livestock was calculated as the number of goats converted from livestock biomass per each herder, assuming the biomass of an individual goat as 50 kg. Both vegetation processes and pastoral events were recorded every 15 days during the pasture season, and the decrease in the number of livestock was recorded once each year at the end of the pasture season. Stochastic simulations were repeated 20 times for each scenario, and the results (i.e., mean changes in DDPA, total biomass, and the number of livestock that each herder held) represented an average value across these 20 replicates.

6.3 Results

This study only focused on the style of pasture mobility and grazing pressure, as a factor of sustainability of the environment and herder livelihood (cf. for the analysis of the effect of other factors). From the results of these spatially explicit simulations (Fig. 6.5), we determined that substantial reductions in livestock and pasture degradation were predicted for both high and moderate levels of grazing in the sedentary system (Fig. 6.6a, b). However, DDPA differed with grazing pressure

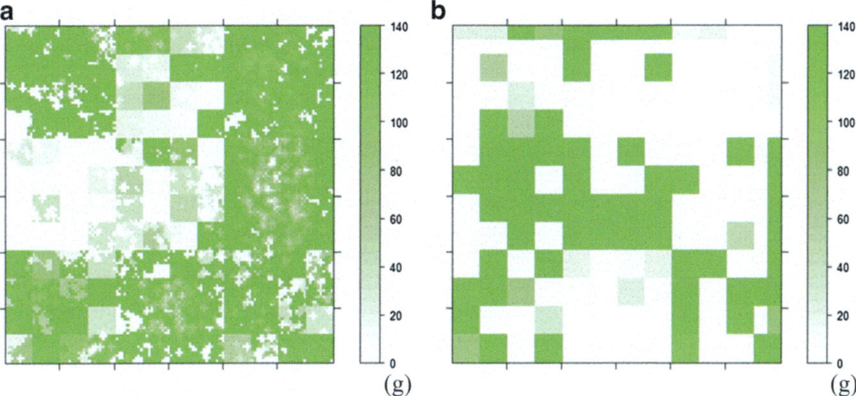

Fig. 6.5 Example patterns of distribution of grassland biomass in pastures at the end of a 30-year simulation in (**a**) a nomadic pastoral system and (**b**) a sedentary pastoral system

under the nomadic system; degradation increased slightly with increasing number of livestock at a moderate grazing pressure but increased gradually and more substantially at high grazing pressure, even despite livestock number reductions (Fig. 6.6a, b).

It was also found that the mean number of livestock the land could support (i.e., the average income of herders in each pastoral system) and its variance under moderate grazing pressure differed depending on the mobility mode, where more livestock could be held on average in the nomadic system compared to the sedentary system (Fig. 6.6c). For example, in typical stochastic simulation runs at moderate grazing pressure, 172 of 529 herders held less than 2 animals of livestock after 30 years of sedentary pasturing while nomads were found to hold at least 309 head after 30 years of nomadic pasturing (Fig. 6.7a, b). A large variance occurred between herders in the number livestock they held (i.e., there was a widening economic gap among herders) in the sedentary pastoral system with a large percentage of the sedentary herders owning less than 2 animals of livestock after 30 years (Fig. 6.7b). In comparison, under high grazing pressure, the number of livestock that could be supported decreased after 30 years, regardless of the mobility mode (Fig. 6.6c), suggesting that pasturing pressure on the real landscape of UB may have already exceeded the land's supporting capacity. Although the number of livestock held by each herder was significantly less under high grazing pressure compared to the moderate grazing scenario (Fig. 6.7), average DDPA under both mobility modes gradually increased (Fig. 6.6a). Thus, even when herders move their livestock from place to place, pastoral systems with high densities of herders will be unsustainable, both environmentally and economically.

After the 30-year simulation of the nomadic mobility system, the total numbers of livestock were similar under moderate and high grazing pressures (Fig. 6.6b), but the numbers of livestock held by each herder (Fig. 6.7a, c) and average

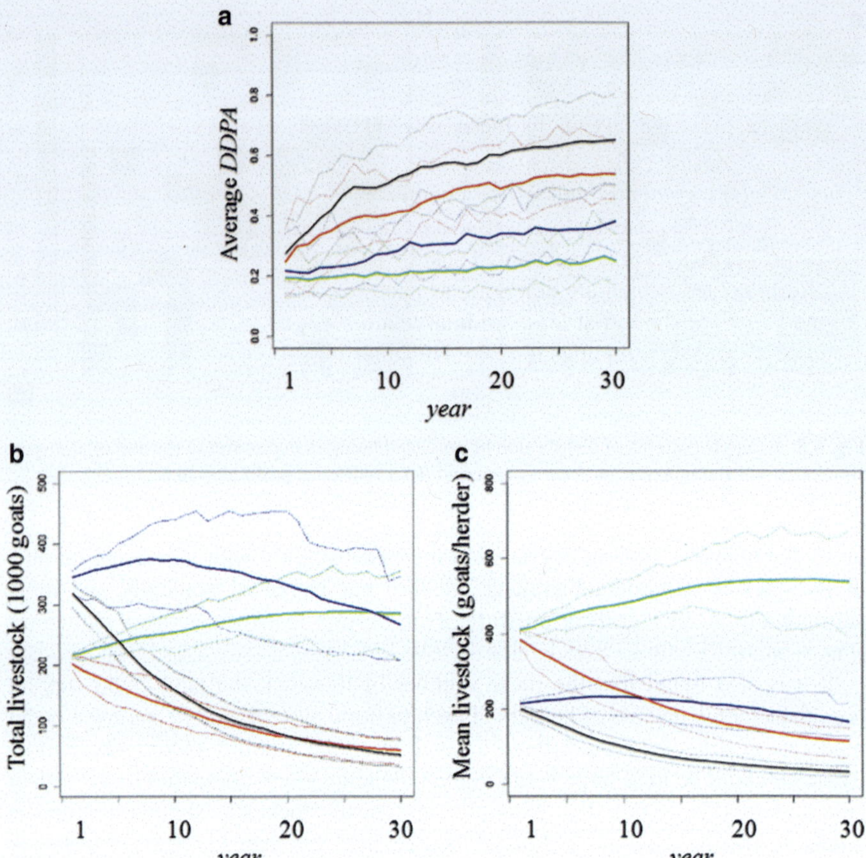

Fig. 6.6 Changes in pastoral systems after 30-year simulations under four scenarios that modeled a nomadic system with moderate grazing pressure (*green line*), a nomadic system with high grazing pressure (*blue line*), a sedentary system with moderate grazing pressure (*red line*), and a sedentary system with high grazing pressure (*black line*). Measured changes include (**a**) the average degree of pasture degradation (DDPA), (**b**) total number of livestock within the pasture area (per 1,000 goats), and (**c**) mean number of livestock per household. *Dotted lines* above and below *solid lines* represent the maximum and minimum values, respectively, which resulted from stochastic simulations

DDPA (Fig. 6.6a) were substantially different between grazing pressures. This comparison suggests that the density of herders in a given area had a greater impact on pasture degradation compared to the number of livestock held by each herder. Thus, limiting the number of herder households may be a more effective strategy for achieving ecological and economic sustainability than limiting the number of livestock owned by individual households. Furthermore, sedentary pastoralism always led to pasture degradation, even at lower grazing levels (Fig. 6.6a), illustrating that mobility is a key factor for environmental and economic sustainability in Mongolian pastoral systems.

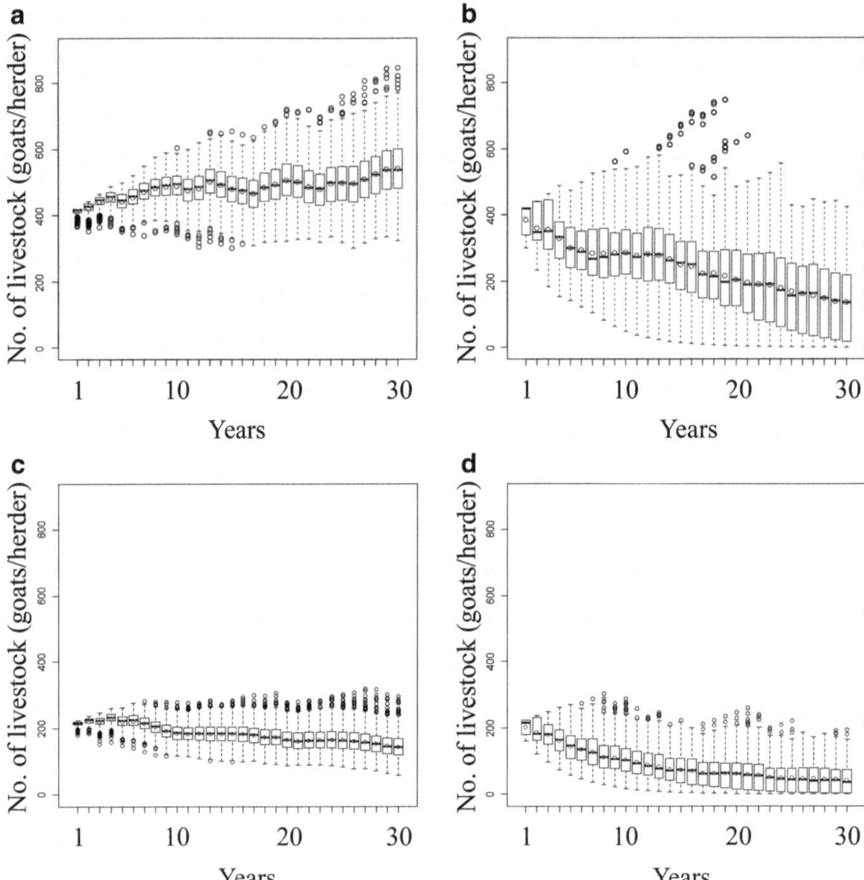

Fig. 6.7 Median (*black bar*) and mean (*red circles*) change in the number of livestock per household through time based on scenarios for (**a**) a nomadic system with moderate grazing pressure, (**b**) a sedentary system with moderate grazing pressure, (**c**) a nomadic system with high grazing pressure, and (**d**) a sedentary system with high grazing pressure. The *white box* represents the range between the first and third quartile of the number of herder livestock after the pasture season each year

6.4 Discussion and Conclusions

In "The Tragedy of the Commons," Garrett Hardin (1968) stated that individuals acting independently and according to one's self-interest would deplete a shared common resource such as a pasture land over time as each individual maximized his consumption of the shared resource over that of other individuals to secure his own profits before exploitation by any other individual. Because multiple individuals attempt to maximize their use of the resource, it is unavoidable that the shared resource is degraded or exhausted (Hardin 1968, 1998). Under Hardin's argument (1968), restriction of unlimited public use of the shared resource and/or the promotion of incentives for sustainable use are required to protect the resource from exhaustion or degradation. On the basis of this logic, the best course of action for preserving pasture lands (i.e., the shared resource) in Mongolia would be to limit their use by giving private ownership to herders, which would reduce land degradation by preventing the overuse of pasture by a large density of herders. Owners would then have an incentive to use sustainable grazing practices to ensure the longevity of their resource. This is the rationale by which the Pasture Law is currently discussed in Mongolia. However, studies using satellite imagery over Russia and China have shown remarkable environmental degradation compared with Mongolia in areas that have shifted from traditional nomadic practices to sedentary agriculture and stock farming through the nationalization or privatization of land (Sneath 1999; Sneath 1998). The results of this study support the prediction that privatization and a shift to sedentary grazing practices would result in environmental degradation and a decline in the economic viability of pasture as an industry. Under sedentary practices, land use by each herder would be fixed, preventing herders from both reacting to spatiotemporal variability in precipitation as risk aversion and reducing their use of grasslands that may be quick to degrade and slow to recover (Fernandez-Gimenez 2006), because herders cannot avoid the pasture field that is less productive and has a low recovery from degradation, and, therefore, they have to overuse, repeatedly such fragile resources.

Ostrom et al. (1999) previously offered similar arguments in response to "The Tragedy of Commons" (Hardin 1968), contending that exhaustion of a shared resource was not always inevitable and could be avoided with the compliance with local rules governing sustainable resource use. However, as indicated by the model in this research, the pastoral system under a scenario continuing with nomadism would not necessarily be designed according to these principles. The results of this study show a certain case when degradation of resources can be avoided, although each herder may try to maximize profit. In Mongolia, nomadic herders would be expected to search for optimal resources and move between high-quality patches to graze livestock (Stephens and Krebs 1986). As a result, grazing pressure would ultimately be guided to resource patches which may be more regenerative, naturally leading to sustainable resource use of shared grasslands. However, if the household density of herders increases or if their mobility decreases within the same area, the opportunity for herders to access and to exploit higher quality, more easily regenerated grassland would decline. In this instance, private ownership would not lead to a reduction in the exhaustion or degradation of a once shared resource. This hypothesis is further supported by the results in this study; nomadic practices maximized the profit of each herder (as measured by the number of livestock owned) and reduced environmental degradation by enabling intermittent and sustainable use of grasslands. To maintain a sustainable pasture system like this one in Mongolia, agricultural and economic policies should allow for the distribution of grazing loads throughout the landscape and limit the number of herders concentrated within a given area.

This study took only herder density at BO and around UB into consideration as a difference between conditions to compare grazing pressure. To evaluate the sustainability of the pasture system in Mongolia, further evaluation is needed using gradual-condition of grazing pressure based on the statistical data of pastoralism outside BO and UB. In addition, the amount of precipitation and its temporal pattern will be another key factor in pastoral system analysis. Precipitation in Mongolia varies greatly, with more precipitation occurring in the north than in the south. Meteorological data must therefore be obtained based on location.

References

Badarch D, Ochirbat B (2002) On sustainable development of cashmere production and goat population in Mongolia. In: Oka H (ed) Mongolian studies at CNEAS. CNEAS Monograph Series 6. Center for Northeast Asian Studies, Tohoku University, Sendai, Japan, pp 187–200

Batjargal Z (1997) Desertification in Mongolia. RALA Rep. No. 200

Bruun O, Odgaard O (1996) Mongolia in transition: old patterns, new challenges. Nordic Institute of Asian Studies: Studies in Asian Topics No. 22, Abingdon, Oxon

Fernandez-Gimenez ME (2001) The effects of livestock privatization on pastoral land use and land tenure in post-socialist Mongolia. Nomadic Peoples 5:49–66

Fernandez-Gimenez ME (2006) Land use and land tenure in Mongolia: a brief history and current issues. In: USDA Forest Service Proceedings RMRS-P-39, pp 30–36. http://www.fs.fed.us/rm/pubs/rmrs_p039/rmrs_p039_030_036.pdf

Hardin G (1968) The tragedy of the commons. Science 162(3859):1243–1248

Hardin G (1998) Extensions of "the tragedy of the commons". Science 280(5364):682–683

Holmes W, Jones JGW (1964) The efficiency of utilization of fresh grass. Grass Forage Sci 19 (1):119–129

Kamimura A (2013) Pastoral mobility and pastureland possession in Mongolia. In: Yamamura N, Fujita N, Maekawa A (eds) The Mongolian ecosystem network: environmental issues under climate and social changes. Springer, Tokyo

Kevin C (2003) Growth and recovery in Mongolia during transition. IMF Working Papers 03/217, International Monetary Fund

Lise W, Hess S, Purev B (2006) Pastureland degradation and poverty among herders in Mongolia: data analysis and game estimation. Ecol Econ 58:350–364

MoFALI (Ministry of Food, Agriculture and Light Industry) (2010) National Mongolian Livestock Program (Draft). Accessed 28 Nov 2013. http://www.mofa.gov.mn/mn/images/stories/busad/mmeng.pdf

Nachinshonhor UG, Jargalsaikhan BL (2013) Seasonal migration and daily herding of Mongolian nomadic people – a case study from the arid steppe of Bayan-Unjuul county, Tov prefecture in Mongolia. In: Yamamura N, Sakai S (eds) Project report: degradation and restoration of ecosystem networks with human activity. Research Institute for Humanity and Nature, Kyoto

Ostrom E (1990) Governing the commons: the evolution of institutions for collective action. Cambridge University Press, New York

Ostrom E, Cox M (2010) Moving beyond panaceas: a multi-tiered diagnostic approach for social-ecological analysis. Environ Conserv 37(4):451–463

Ostrom E, Burger J, Field CB, Norgaard RB, Policansky D (1999) Revisiting the commons: local lessons, global challenges. Science 284(5412):278–282

Pomfret R (2000) Transition and democracy in Mongolia. Eur Asia Stud 52:149–160

Sneath D (1998) State policy and pasture degradation in inner Asia. Science 28(5380):1147–1148

Sneath D (1999) Spatial mobility and inner Asian pastoralism. In: Humphrey C, Sneath D (eds) The end of nomadism? White Horse Press, Isle of Harris, pp 218–277

Stephens DW, Krebs JR (1986) Foraging theory. Princeton University Press, Princeton

Yamamura N (2013) Ecosystem network as social-ecological systems. In: Yamamura N, Fujita N, Maekawa A (eds) The Mongolian ecosystem network: environmental issues under climate and social changes. Springer, Tokyo

Chapter 7
Trajectories of Adaptation: A Retrospectus for Future Dynamics

Donald R. Nelson, Francisco de Assis de Souza Filho, Timothy J. Finan, and Susana Ferreira

Abstract Sustainable adaptation to climate change needs to be assessed beyond the present time and location to include the way that current forms of adaptation might influence future response options. An analysis of past dynamics of adaptation, what we call "trajectories," might hold the key to understanding how the adaptive outcomes of past responses to climate stress constrain or open avenues to future adaptation. Adaptation research often focuses on particular actions, technologies, or institutions which may positively influence these relationships in order to build resilience and reduce vulnerability. However, relationships are complex and often behave in unexpected ways. There is no simple cause and effect, but rather actions are modified and transmitted through a web of linkages and feedbacks that are both physical and social. This complexity challenges our ability to predict the outcome of particular actions and there remain gaps in the understanding of system interactions that would permit a more accurate assessment of future development trajectories. The work presented here is an analysis of change in the climate vulnerability of dryland farmers in Northeast Brazil over four decades. The analytical framework, which links biophysical characteristics with a socio-economic context and indicators, permits an analysis that captures the dynamic relationship of adaptive capacities and consequent changes in vulnerability. The analysis of trajectories provides a foundation for future assumptions about human behavior and the relationship with the environment.

D.R. Nelson (✉)
Department of Anthropology, University of Georgia, Athens, GA, USA
e-mail: dnelson@uga.edu

F.d.A. de Souza Filho
Department of Hydrological and Environmental Engineering,
Universidade Federal do Ceará, Fortaleza, Brazil

T.J. Finan
Bureau of Applied Research in Anthropology, University of Arizona, Tucson, AZ, USA

S. Ferreira
Department of Agricultural and Applied Economics, University of Georgia, Athens, GA, USA

S. Sakai and C. Umetsu (eds.), *Social–Ecological Systems in Transition*, Global Environmental Studies, DOI 10.1007/978-4-431-54910-9_7, © Springer Japan 2014

Keywords Adaptive capacity • Drought index • Dryland farming • Governance • Rainfall variability • Scenarios

7.1 Introduction

Adaptation describes an on-going and continuous process of interaction between human and natural systems. Natural system dynamics, such as global warming and climate change, introduce stressors into human system configurations which affect such human system outcomes as well-being, quality of life, or even survival. It is well-known that human system dynamics involving the introduction of technology and ecosystem also demands the alteration of natural system characteristics. Vulnerability describes the "success" of human system response to natural stressors and is assessed in terms of the magnitude of natural system variability, the sensitivity of the human systems to the impacts of this variability, and the capacity of the systems to absorb or recover from these impacts. Human system response outcomes which absorb natural stressor impacts and result in recovery which does not alter the "integrity" of the human system are said to indicate low levels of vulnerability and strong resilience (Folke 2006).

Vulnerability, however, is a shifting and not a static quality of human systems. As Walker et al. (Walker et al. 2004) have theorized, human systems "learn" in the sense that the very process of adaptation can result in changes in human response and in the nature of the natural system interaction, thus reducing vulnerability. On the other hand, systems which fail to learn may see their vulnerability increase. Furthermore, while responses to perceived or actual climatic changes may reduce current vulnerability, they may also reduce capacities to respond to future events (Barnett and O'Neill 2010; Fazey et al. 2011). This chapter seeks to identify this process of system learning as adaptation through time. We suggest here that the adaptation process reveals itself in past social-ecological system interactions and the understanding of past responses, particularly learning responses, provides an analogy for future adaptation. These changes (or not) in vulnerability through time, the outcome of natural and human system interaction, are characterized as "trajectories of adaptation," observable, even measurable, movements in the interaction of natural and human systems.

We assume here that such trajectories do not reflect random events, but the outcomes of specific efforts on the part of individuals, households, livelihood systems, and governments to adjust to variability in natural systems. The learning taking place over time may be manifest in the widespread adoption of new technological options, increased economic investment, a government policy set which neutralizes the impacts of natural system variability, or even improvements in governance and co-management strategies. Significant research is invested to identify the types of changes positively influencing levels of vulnerability and resilience (Conway and Schipper 2011; Gotham and Campanella 2011; Hisali et al. 2011; Smit and Wandel 2006). The lessons learned from this analysis of trajectories provide the inspiration for future strategies in adaptation to natural system change.

Trajectories trace the relationship of humans with the natural environment over time and they allow comparison of similar systems across space. This approach helps identify what is already working to reduce sensitivity, and what has made a positive difference in the lives of people. There is a need to recognize what has worked in a particular instance to reduce sensitivity, or, conversely, what may have aggravated sensitivity. Too often, however, vulnerability assessments are designed to identify deficiency rather than adequacy. This gap underscores the contribution of an analysis of trajectories. Additionally, adaptations need to be culturally and contextually sensitive, and there are few out-of-the-box adaptations which will work in all contexts. Critical knowledge for adaptation is contingent on the relevance to particular societies and cultures. Knowledge to help prepare for and respond to large-scale change needs to be situated within local contexts and understandings of society and the environment. Local and place-based characteristics of adaptation include unique cultural logics which guide human action and contribute to adaptation outcomes (Colombi and Smith 2012; Adger et al. 2012). Humans are able to evaluate information from the past, speculate about the future, and incorporate values and norms into decision-making processes. As a result, interventions in similar ecological systems may have dramatically different outcomes as a result of the way interventions are translated through cultural systems.

In analyzing particular trajectories we do not suggest the future in each location is determined by the past, but rather there is an inherent logic related to human response characteristics, both in the private sphere as well as in the public and policy spheres. Past behavior provides insight into the future, not because behaviors are deterministic, but because people tend to respond to challenges based on past experiences and within a range of choices determined by socio-political structures which are slow to change (Scheffer 2009). There is value in identifying common characteristics and analyzing patterns in systems across the globe. However, the ability to respond proactively to current challenges will need to draw from the rich and detailed diversity of experience and adaptation resident within populations around the world.

An understanding of the local nature of adaptation permits a more informed vision of possible futures. Similar to the need for climate models to be calibrated through backcasting, models used to forecast or develop scenarios of future social-ecological states need to be grounded in an empirical understanding of past behaviors and change. The ability to understand the trajectories of social–ecological systems, in the light of climate vulnerabilities, provides an indication of where a system is heading (Fazey et al. 2011) and where leverage points exist which may help guide the future trajectory. Unlike methods which capture only a snapshot of current vulnerability, methods incorporating past trajectories have a much stronger basis for developing plausible future scenarios (Thompson et al. 2012).

Our method to identify and analyze trajectories is straightforward. As described below, we have used time-series and cross-sectional data to identify natural system variability at the local level. Departing from the hypothesis that natural system variability has an impact on human systems, we have created impact measures sensitive to natural system variation, then analyzed this relationship over a period of four decades to identify locally the evidence for adaptation.

The objective of this chapter is to contribute to the growing literature focused on sustainable, long-term adaptations. To this end, we present an analytical framework designed to assess historical changes in the climate vulnerabilities of populations, in order to inform decisions about immediate and future adaptations. We illustrate the framework and the analytical contributions through a discussion of 40 years of public and private investments in drought adaptation in Northeast Brazil. In the section below we discuss the concept of trajectories and their potential contribution to adaptation studies. This is followed by a description of the research area, including a look at the environmental variability in the region and the importance of a focus on the local nature of adaptation. We then detail the way in which our model integrates biophysical indicators with socio-economic indicators to provide a perspective on the relationships between system components. We describe model outputs and how they contribute to the identification of trajectories of adaptation. Finally, we outline what we believe to be the primary contribution of a trajectory framework and how the results can contribute to a more informed, planned process of adaptation.

7.2 Trajectories of Adaptation

Vulnerabilities and adaptive responses can be documented as they change over time. The concept of trajectories provides a way to explore these changes in systems. We state that the trajectory of a system provides an invaluable perspective to prepare for current and future challenges. The analysis of trajectories grounds future assumptions about human behavior and the relationship with the environment. To predict plausible future directions, a basic understanding of past trajectories and change is required. The ability to predict the future location of a celestial body, for example, requires more than simply knowing its current location. It is necessary to describe the trajectory, to determine the forces acting on the body, and to understand how and where these forces will direct the body. The same perspective holds true for social–ecological systems. Trajectories inform an understanding of how contemporary situations of vulnerability and resilience arose. Importantly, they also provide a forward-looking view that anticipates the dynamic relationship between human actions and uncertain change. For example, in order provide insight into possible future scenarios and develop appropriate public investment strategies, methods, such as participatory timelines that capture relationships within social-ecological systems are used to elicit changes in vulnerability over time (Enfors et al. 2008).

The concept of trajectories implies three fundamental qualities related to planning and adaptation. First, trajectories describe pathways. In common discourse we often think about individuals, populations, and systems as following pathways.

The pathways may take on normative characteristics. For example, an individual can be on a path to betterment or self-destruction. Modernization proponents claim that investments and technology will put populations on a pathway to increased wealth, happiness, and consumption (e.g., Rostow 1990). The same concept is relevant for social–ecological systems. Vulnerability assessments frequently offer snapshots of a particular time and place, and, if not historical, are static and detached. The concept of pathways helps to conceptualize the continuity of systems through time and link past actions with current states of being.

Thus, trajectories also suggest historical depth. Contemporary literature on adaptation recognizes the importance of path dependency and legacies and how these historical artifacts contribute to the shaping of decision-making processes. Past decisions have the potential to limit our vision and response options. Pathways can be good or bad, but the human dimensions literature suggests that, to be able to respond to unexpected change, actors need to have access to a diversity of possible response options (Folke 2006). There is thus a forward-looking focus which seeks to understand how to break out of path dependencies and to reduce the influence of physical and social legacies. Historical depth also provides perspective on how a particular system came to be in its current state and helps identify current trajectories. It provides insight into why a particular population is vulnerable and the historical contingencies that created the context.

The third quality of a trajectory relates to inflection points. Inflection points mark locations of significant change, in which the directionality and curvature of a trajectory changes. They are the loci of change that provide insight into how trajectories are altered and they call attention to the set of factors necessary to induce change. Reflecting on the metaphor of a celestial body, analysis of inflection points can provide an empirical understanding of the forces at work at a given point responsible for controlling the direction of a system trajectory. If the slope of a particular trajectory represents the sensitivity of a system to climate perturbations over time, a constant slope indicates that sensitivity hasn't changed. However, at the point of inflection the slope changes and increases or decreases in sensitivity can be noted and tracked. This knowledge can illuminate the inner workings of a system and can be used to identify leverage points to initiate the desired change. Here, we concern ourselves with a focus on pathways and historical depth.

7.3 Adaptation on the Ground

The public and private adaptations occurring in the study region and elsewhere around the world are responses to perceived hazards and are intended to reduce vulnerability (Nelson et al. 2007). Here we consider vulnerability as the level of susceptibility of an analytical unit to a particular type of event in which the susceptibility is described by the characteristics of exposure, sensitivity, and response capacities (Adger 2006). Populations better adapted to drought demonstrate lower levels of vulnerability and, in our framework, changes in levels of sensitivity serve as proxies for changes in responses and levels of adaptation.

Ultimately, climate adaptation is constrained and facilitated by local (and finally the household level) parameters and phenomena. Sensitivity to climate perturbations is a function of the characteristics of the perturbations themselves, as well as social-ecological conditions and relationships. Northeast Brazil is well represented in the tropical drylands literature, particularly in relation to drought sensitivity. Much of the region falls within what is known as the *Polígono das Secas*, or Drought Polygon. The literature on the Drought Polygon frequently cites familiar indicators of biophysical characteristics including scarce and variable precipitation, high temperatures and solar radiation, high rates of evapotranspiration, and low soil fertility. On the social side, emphasis is directed towards levels of absolute poverty, measures of income inequality, out-migration, clientelism, and low education rates.

While perhaps representative in general, these characterizations gloss over the nuanced social-ecological heterogeneity across the landscape. Vulnerability is defined by the intersection of wider contextual variables with very local and specific characteristic sets through which stresses are propagated. Thus even within a given region, the experience of drought will vary. Trajectories are able to document the variety of relationships at regional and very local scales. The purpose of this chapter is not to detail this heterogeneity across Ceará, a state that is approximately 149,000 km^2. However, a brief description of the region, in particular the variability in rainfall and the availability of soil moisture, illustrates the claim that risk management strategies and adaptation must account for the local context.

Public responses to drought have a long history in Northeast Brazil. The vulnerability of the population came to public light during a severe regional drought in 1877–1879 in which hundreds of thousands of people died. In response, the federal government created an institution to "combat" drought. Its primary mandate was to increase water surface storage across the region through the construction of large and small dams. These water storage activities continue today. Public responses have passed through several phases since the first dams were built, all of which have approached the question of vulnerability from a technological perspective (Finan and Nelson 2001). These efforts include a focus on increasing agricultural productivity through the introduction of drought-tolerant crops, the development of large-scale irrigation projects, and attempts to modernize agriculture through increased mechanization. Yet, because the majority of the rural population did not have access to these programs, their impact was minimal (Frota and Aragão 1985).

Subsistence agriculture continues to figure prominently in the livelihood strategies of rural households. Beans and maize are the principal crops and production is based almost entirely on available rainfall. The highly variable nature of rainfall in the region translates into high-risk agriculture. Farmers have developed a number of household-specific adaptations to help distribute and manage risk over time and space. These strategies include planting a diversified set of bean and maize cultivars and planting across different fields in order to distribute exposure to weather events. One of the most critical decisions that a farmer makes is when to seed in the ground and there are numerous planting strategies to spread the risk of crop failure, including replanting.

The single rainy season can start as early as the beginning of December and can last into May. Most farmers will prepare their land in anticipation of the onset of the rainy season and wait to sow their fields until there is sufficient soil moisture to guarantee germination. However, if a farmer prepares too soon, labor is wasted because weeds and shrubs will take over the field. If he waits too long the best planting opportunities may be missed. Thus the decision of when and how to plant is a highly strategic and stressful decision and is the subject of constant speculation throughout the year.

In order to minimize the chance of total loss, few farmers plant their entire crop at once. This strategy is in response to the phenomenon called the *veranico*. A veranico is a short period (10–12 days) with little to no rain which can significantly impact agricultural productivity and often wipes out the first planting. It is common to have hard, plentiful rain, sufficient to force germination, which is then followed by a veranico. The challenge of a farmer is to assess not only the current soil moisture, but also the chance that this was an isolated rainfall. Thus, planting times are often staggered, either across fields or within an individual field. The following figures document the range of variation across space and through time for the 184 municípios in Ceará and highlight the decision-making context in which farmers must make critical agricultural decisions. The figures present different analytical lenses to explore rainfall and soil moisture variation as it relates to subsistence maize and bean production.

Figure 7.1a, b presents histograms which summarize the best planting date for beans in two municípios (Fig. 7.2a, b summarizes the best planting dates for maize), and which capture some of the risk management challenges faced by the farmer. The process to determine "the best planting date," depicted in the figures, is detailed in the section 7.4. The difference in the range of best planting dates of both crops is statistically significant, but, more importantly, it is significant from a risk management perspective. Maize farmers in Potiretama (Fig. 7.1a) and in Capistrano (Fig. 7.2a) have a much larger range of possible "best planting dates" than do subsistence producers in Chaval (Fig. 7.1b) and Barbalha (Fig. 7.2b). In Chaval, heavy rains in December have always been isolated rains. Even if the soil is sufficiently moist to guarantee germination, these rains are followed by veranicos. The Chaval farmers can rest assured that they don't need to prepare their fields in November or plant in December. The Potiretama farmer, however, has no such historical assurances. During the last 20 years their best planting dates have fallen across a range of 5 months. These differences have significant implications for individual farming strategies, but also for the other ways in which farmers seek to manage livelihood risks.

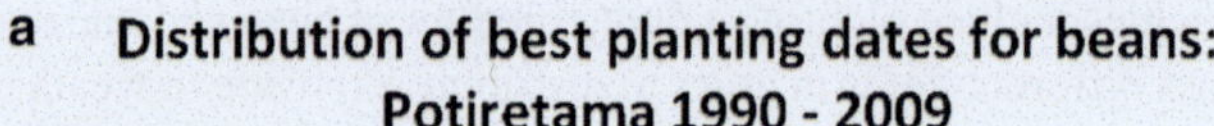

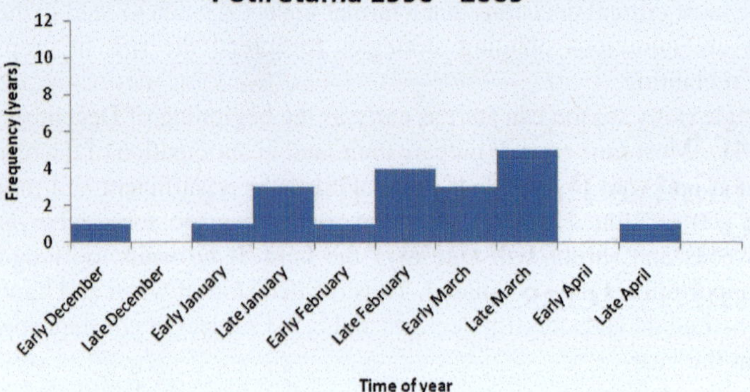

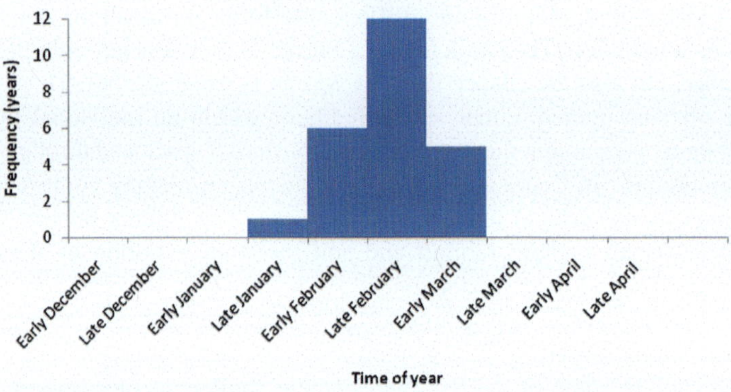

Fig. 7.1 (**a**) Soil moisture variation and bean productivity—Potiretama. (**b**) Soil moisture variation and bean productivity—Chaval

7.4 Trajectory Model

Two data sets were constructed in order to assess the relationship between rainfall and socio-economic indicators and describe the trajectories. The first documents the annual variation of soil moisture availability in relation to the needs of maize and beans. We gathered daily rainfall values across the state of Ceará for the years 1973–2009. The data was registered in a total of 193 gauges—although some gauges came online after 1973. Each gauge also has soil attribute data necessary to create a daily soil moisture model (Fig. 7.3).

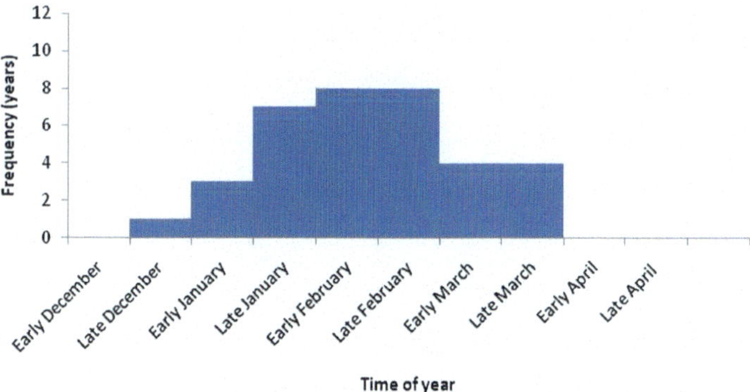

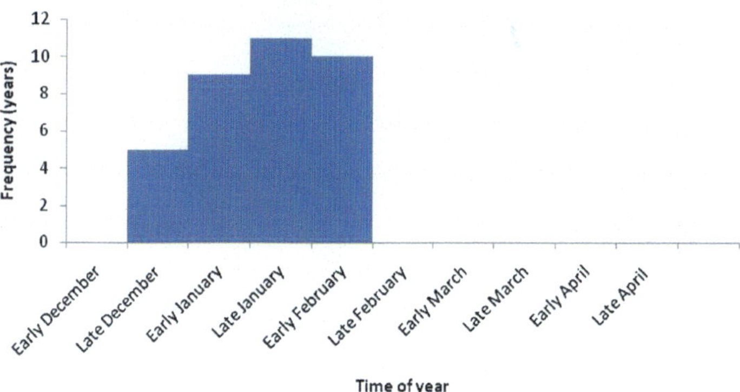

Fig. 7.2 (**a**) Soil moisture variation and maize productivity—Capistrano. (**b**) Soil moisture variation and maize productivity—Barbalha

$$US_i = US_{i-1} + Pr_i + AC_i + I_i + ES_i + ETR_i + PP_i \tag{7.1}$$

where US_i is soil humidity at the end of day i; US_{i-1} is soil humidity at the end of the previous day, $i-1$; Pr_i is rainfall on day i; AC_i is capillary uptake for day i; I_i is irrigation for day i; ES_i is surface runoff for day i; ETR_i is real evapotranspiration for day i; and PP_i is deep percolation for day i.

This model was then used to estimate potential crop loss for each of the years in the data series for each of the rain gauges. The methodology for estimating crop productivity followed FAO guidelines in which relative yield reduction is related to the corresponding relative reduction in evapotranspiration (Allen et al. 1998; Doorenbos et al. 1979). Productivity loss is described based on the function of the evapotranspiration deficit during the phenological cycle of the crops in question

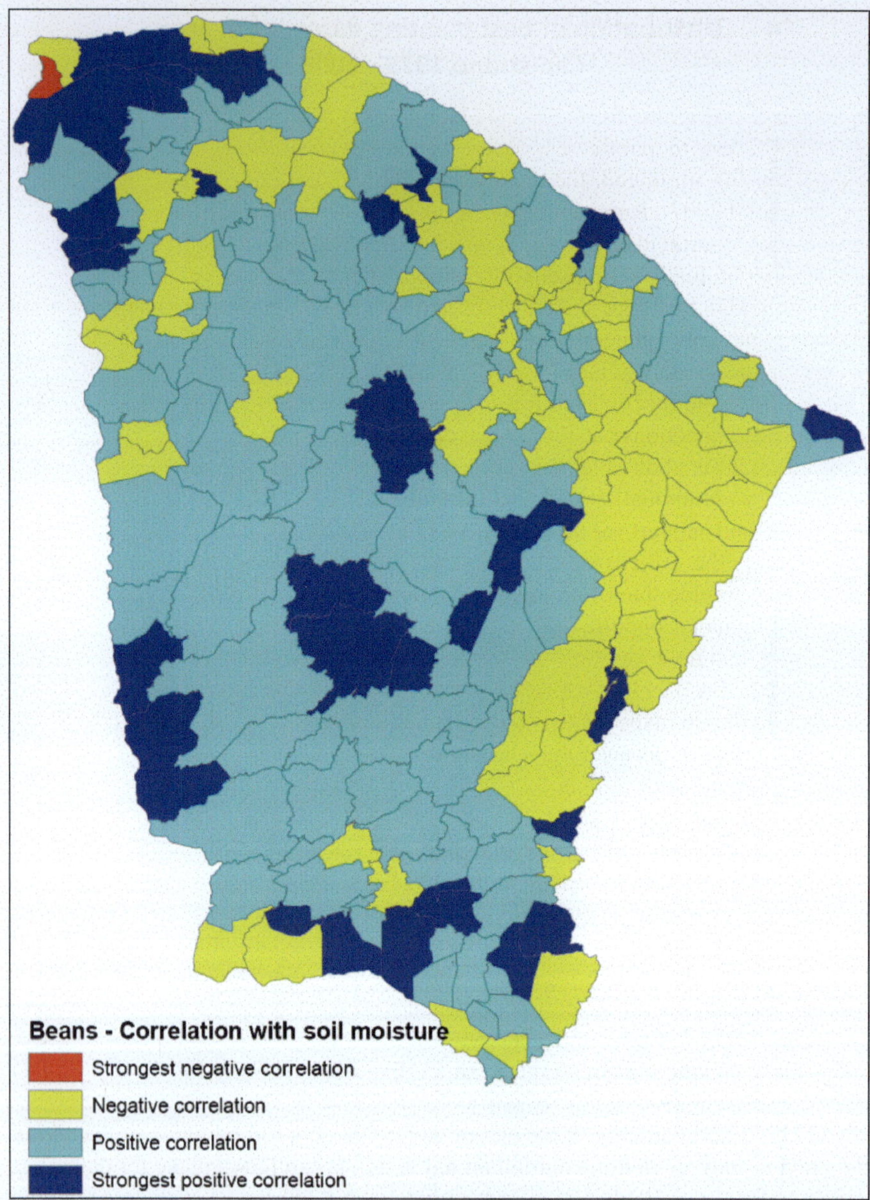

Fig. 7.3 Nature index β-values for predicting bean productivity

shown in (7.2) (Doorenbos et al. 1979). Model simulations were run for maize and beans from December 1st through April 30th. The models do not consider any type of soil management or other adaptations, but only potential evapotranspiration of the crop, the type of soil, and the timing and amount of rainfall. For each simulated day the model identified the maximum productivity for maize and beans. The model output does not represent actual productivity, but rather the estimated maximum productivity if an agriculturalist had planted on any given day. The model output is referred to as the Nature Index and represents the highest productivity for a given year. This number provided the best estimates in the regressions described below. The first maximum value in a year is what we refer to as "the best planting date" (see Figs. 7.1 and 7.2).

$$\left(1 - \frac{Y_{\mathrm{a}}}{Y_{\mathrm{m}}}\right) = k_{\mathrm{y}}\left(1 - \frac{\mathrm{ET_R}}{\mathrm{ET_m}}\right) \tag{7.2}$$

where Y_{m} and Y_{a} are the maximum and actual yields; $\mathrm{ET_m}$ and $\mathrm{ET_R}$ are the maximum and actual evapotranspiration; and k_{y} is a yield response factor representing the effect of a reduction in evapotranspiration on yield losses.

The daily model results were then extrapolated to município level. The município level productivity was calculated using the weighted average of productivity calculated at each rain gauge (Prod_i) and the area of influence of each gauge (A_i), as described in (7.3). The area of influence was calculated using the Thiessen method.

$$\mathrm{Prod_m} = \frac{\sum (A_i \mathrm{Prod}_i)}{A} \tag{7.3}$$

A second data set was developed that includes socioeconomic factors. These are publicly available, município-level time series data. The variables include actual production and productivity levels for maize and beans, município tax revenue, educational enrolment, infant mortality, emergency declarations, and livestock production. For the current analysis, a linear regression model used the nature index to predict each of the socioeconomic variables. The parsimonious model is designed to highlight relationships between each of the individual variables with the nature index. Two equations were used. Equation (7.4) explores variation in climate sensitivity across the state, and (7.5) is used to evaluate annual performance of each município. Outputs are discussed in the following section.

$$Y_{jt} = \alpha + \beta * f\left(\text{nature index}_{jt}\right) + \mu_j + \varepsilon_{jt} \tag{7.4}$$

where Y_{jt} is the socio-economic indicator in município j, year t; β is the expected change of the indicator for a one-unit change in the nature index, *ceteris paribus*; μ_j is the município-level fixed effects; and ε_{jt} is an error term.

$$Y_t = \alpha + \beta * f(\text{nature index}_t) + \varepsilon_t \tag{7.5}$$

where Y_t is the socio-economic indicator, year t; β is the expected change of the indicator for a one-unit change in the nature index, *ceteris paribus*; and ε_t is an error term.

7.5 Trajectories Revealed

Each of the indicators is hypothesized to be sensitive to changes in rainfall. For example, we hypothesize that bean productivity declines in years with a lower nature index and increases in years with a higher nature index. Where this is not the case, we anticipate human adaptations. Figure 7.3 provides an example of the relationship between the nature index and bean productivity. The map reports the β values at the município level based on data from 1975 to 2009. The darkest municípios are those with the strongest positive correlation between soil moisture variability and bean productivity. As we expect, productivity in these municípios tends to follow the rain. However, this is not the case across the entire state. The lightest colored municípios are those in which there is not a positive correlation and which are less sensitive to rainfall variation. In these municípios the inter-annual variation of rainfall and soil moisture does not explain the variation in productivity. For people familiar with Ceará, the map makes intuitive sense. Many of the light colored municípios overlap with the Jaguaribe River. The Jaguaribe was dammed early in the twentieth century and has since be the focus of intensive investment in irrigation infrastructure. Note that this map does not represent rainfall or soil moisture variability, but rather it represents the interaction of rainfall variability with farming technologies, strategies, and adaptations. The lighter municípios are those in which sensitivity to climate variation has been reduced through human action.

Many vulnerability assessments demonstrate current levels of sensitivity, similar to the information documented in Fig. 7.3. While this knowledge is important and can be used to help develop priority areas for policy intervention, it provides little information in terms of pathways or historical depth. Here we argue that the addition of trajectories provides a value-added perspective for short- and long-term planning. For example, a município that is currently vulnerable to drought, but which has had reduced vulnerability over the last four decades, has different needs compared to those of a município that is also currently vulnerable but has become more vulnerable over the last 40 years. Figure 7.3 demonstrates the heterogeneity of current sensitivity across the state. However, due to our long experience with the region, we also know that sensitivity has changed differentially across the years of this study. Thus a município showing very high sensitivity in 1975 may have lower levels of sensitivity in 2009. Although the rainfall patterns and the nature index may be similar in 1975 and 2009, the losses incurred in a particular município may be significantly different depending on the adaptations implemented during this time period. Figure 7.3 is a static view of sensitivity and does not capture the dynamics. However, the ability to explain this type of change provides invaluable information to individuals responsible for investing public and private funds for adaptation and development.

Município level trajectories provide insight into where and how vulnerability has changed during the last four decades. A linear regression model fits a line across each of the years of the study to describe the relationship of bean productivity with

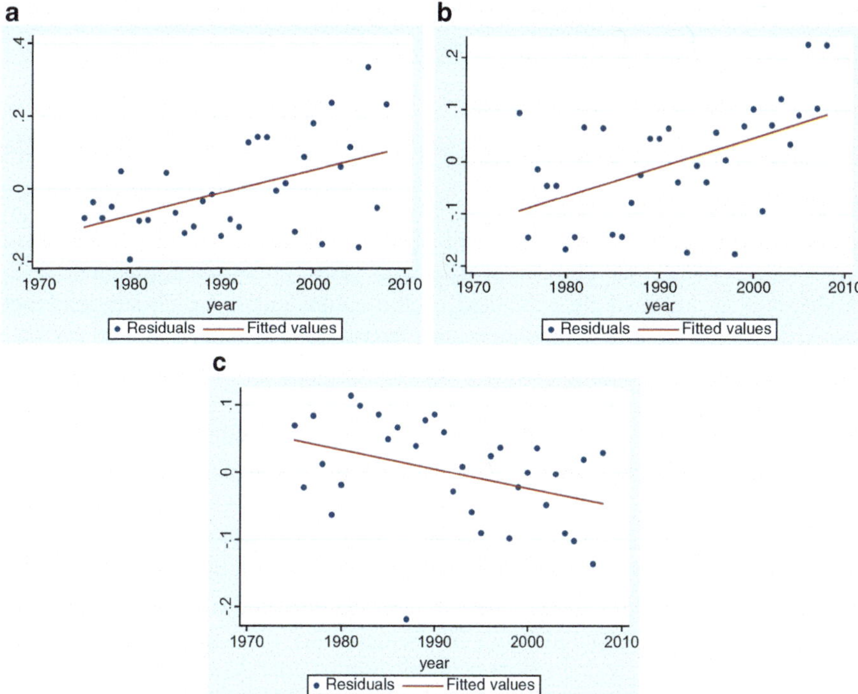

Fig. 7.4 (**a**) Acopiara error term in the nature index—bean productivity regression, plotted against time. (**b**) Baturite—error term in the nature index—bean productivity regression, plotted against time. (**c**) Frexerinha—error term in the nature index—bean productivity regression, plotted against time

the nature index in a single snapshot. Productivity describes the ratio of the production output (beans, kg) per unit of land (ha) and is therefore normalized over the years. In order to explore changes in the relationship throughout the study period, we plotted the error terms against time. This provides an indication of the strength of the relationship described by the regression line for each of the years. There are roughly three categories of relationships that emerge from this type of analysis: municípios in which the magnitude of the error term varied uniformly over time; those in which the magnitude decreased through time; and those in which it increased. The error term represents influences on bean productivity that are not explained by the nature index.

Figure 7.4a–c shows the output of the plots of the error terms against time for three municípios. Two of the municípios (Fig. 7.4a, b) demonstrate strong positive relationships between the error term and the passage of time ($p < 0.05$). We interpret these findings to suggest that, over time, the bean productivity in these two municípios has become less sensitive to the nature index. This indicates that other factors begin to contribute more significantly to the level of bean productivity and that climate sensitivity has decreased over the last several decades. In the third

Table 7.1 Parameter estimates for the relationship between error term and time

Figure	Coef.	Std. Error	t	$P>\|t\|$	95 % Conf. Interval	
Figure 7.4a: Acopiara	0.006	0.002	3.08	0.004	0.002	0.01
Figure 7.4b: Baturite	0.006	0.002	3.33	0.002	0.002	0.01
Figure 7.4c: Frecherinha	0.003	0.001	−2.24	0.032	−0.005	−0.000

município the relationship is reversed. There is a strong negative correlation ($p<0.05$) between the error term and time. This finding suggests that bean productivity is increasingly explained by the nature index, which suggests that sensitivity to the environment is increasing over time (Table 7.1).

7.6 Conclusion

Our ability to prepare for the future, which is informed by an increased understanding of potential climate change and the recognition of the embedded quality of humans in nature, requires new types of information and analyses. Specifically, there is a need to understand better the vulnerability of a system in relation to variability and how the vulnerability changes through time. Vulnerability indicators need to be robust across outcomes. The weather events projected as a result of a changing climate within semi-arid regions will have a variety of impacts, which include food and economic security, morbidity, and others. Effective adaptation and development policy will be based on identifying determinants sensitive to policy changes and robust across the range of outcomes and which are proven to have had a positive impact in the local context. The data presented above focuses on only one outcome, but the analysis can be expanded to explore a suite of outcomes to assess better the trajectory of the system. This chapter presents an analytical framework for providing the type of information that policy makers will require. It also presents an empirical example of how to operationalize the framework.

Our findings demonstrate the importance of local context in determining vulnerability. All 184 municípios in Ceará are regulated by the same federal and state legislation, but the way in which policies and resources are operationalized varies, and there is significant heterogeneity in social indicators and institutions across the state. It is important to emphasize that the trajectories we explore are not measures of development. Indicators such as the Human Development Index already exist. Rather, trajectories are a measure of the sensitivity to rainfall variation, a natural characteristic of the region which continues to play an influential role in the lives of people. As suggested by work in Ethiopia, reducing sensitivity can positively influence poverty and growth rates (World Bank 2005). The analysis identifies municípios in the state which have been able to reduce sensitivity across a variety of socio-economic indicators. These municípios provide a source of experience which can inform and subsidize state-level policy. Although the preliminary analysis considers each variable in isolation, there is a need to identify how the determinants of vulnerability co-vary over time. Adaptive capacities are interrelated. Due to covariance, there are tradeoffs associated with adaptation decisions that make it impossible to maximize all adaptive capacities. Further analysis will seek to explain covariance over time in order to provide insight into the tradeoffs in investments and identify leverage points for maximizing investments.

A critical step in the analysis is the transition from the identification of patterns to the identification and explanation of the causes of changes in vulnerability. The analysis of secondary data provides a description of the spatial distribution of current vulnerability and how that vulnerability has changed throughout the decades. However, it doesn't serve to explain what underlies the trajectories of vulnerability. Future work will be dedicated to uncovering the particular sets of adaptations that changed the relationships between society and the environment in some of the municípios demonstrating the most compelling changes. The value of this on-the-ground work is that it will be possible to identify specific actions and activities which have led to reduced sensitivity or to increased sensitivity over 40 years.

This chapter has advanced the concept of "trajectories" as a way of interpreting past evidence of changes in adaptation as a possible analogue for future change under increasing natural system variability. The central message here is that changes in the relationships between natural and human systems occur at a local level and are influenced by local contextual factors. The time-series analysis also identifies points in time where trajectories are reconfigured, where "inflection" points caused either by the introduction of new technologies or the impact of new policies redirect the slope of the trajectories in some places. It is assumed that the understanding of these inflection points and their causes will help us to understand the available options under climate change scenarios for the Northeast. While these causes are beyond the scope of this chapter, the research agenda is clear.

References

Adger WN (2006) Vulnerability. Glob Environ Chang 16(3):268–281. doi:10.1016/j.gloenvcha.2006.02.006

Adger WN, Barnett J, Brown K, Marshall N, O'Brien K (2012) Cultural dimensions of climate change impacts and adaptation. Nat Clim Chang 3(2):112–117. doi:10.1038/nclimate1666

Allen R, Pereira L, Raes D, Smith M (1998) Crop evapotranspiration (guidelines for computing crop water requirements). FAO, Rome, 15 pp

Barnett J, O'Neill S (2010) Maladaptation. Glob Environ Chang 20(2):211–213. doi:10.1016/j.gloenvcha.2009.11.004

Colombi BJ, Smith CL (2012) Adaptive capacity as cultural practice. E&S 17(4). doi:10.5751/ES-05242-170413

Conway D, Schipper ELF (2011) Adaptation to climate change in Africa: challenges and opportunities identified from Ethiopia. Glob Environ Chang 21(1):227–237. doi:10.1016/j.gloenvcha.2010.07.013

Doorenbos J, Kassam AH, Bentvelsen CIM (1979) Yield response to water. FAO irrigation and drainage paper, vol 33. Food and Agriculture Organization of the United Nations, Rome

Enfors EI, Gordon LJ, Peterson Garry D, Bossio D (2008) Making investments in dryland development work: participatory scenario planning in the Makanya catchment, Tanzania. Ecol Soc 13(2):42

Fazey I, Pettorelli N, Kenter J, Wagatora D, Schuett D (2011) Maladaptive trajectories of change in Makira, Solomon Islands. Glob Environ Chang 21(4):1275–1289. doi:10.1016/j.gloenvcha.2011.07.006

Finan TJ, Nelson DR (2001) Making rain, making roads, making do: public and private adaptations to drought in Ceará, Northeast Brazil. Climate Res 19:97–108

Folke C (2006) Resilience: the emergence of a perspective for social–ecological systems analyses. Glob Environ Chang 16(3):253–267. doi:10.1016/j.gloenvcha.2006.04.002

Frota GPd, Aragão L (1985) Documentação Oral e a Temática da Seca. Centro Gráfico, Senado Federal, Brasília

Gotham KF, Campanella R (2011) Coupled vulnerability and resilience: the dynamics of cross-scale interactions in Post-Katrina New Orleans. E&S 16(3). doi:10.5751/ES-04292-160312

Hisali E, Birungi P, Buyinza F (2011) Adaptation to climate change in Uganda: evidence from micro level data. Glob Environ Chang 21(4):1245–1261. doi:10.1016/j.gloenvcha.2011.07.005

Nelson DR, Adger WN, Brown K (2007) Adaptation to environmental change: contributions of a resilience framework. Ann Rev Environ Res 32:395–419. doi:10.1146/annurev.energy.32.051807.090348

Rostow WW (1990) The stages of economic growth. A non-communist manifesto, 3rd edn. Cambridge University Press, Cambridge/New York

Scheffer M (2009) Critical transitions in nature and society. Princeton University Press, Princeton

Smit B, Wandel J (2006) Adaptation, adaptive capacity and vulnerability. Glob Environ Chang 16 (3):282–292. doi:10.1016/j.gloenvcha.2006.03.008

Thompson JR, Wiek A, Swanson FJ, Carpenter SR, Fresco N, Hollingsworth T, Spies TA, Foster DR (2012) Scenario studies as a synthetic and integrative research activity for long-term ecological research. Biosci 62(4):367–376. doi:10.1525/bio.2012.62.4.8

Walker BH, Holling CS, Carpenter SR, Kinzig A (2004) Resilience, adaptability and transformability in social-ecological systems. Ecol Soc 9(2):5

World Bank (2005) A strategy to stimulate and balance growth in Ethiopia, Washington, DC

Part IV
Resilience of Social Systems

Chapter 8
Ingredients for Social-Ecological Resilience, Poverty Traps, and Adaptive Social Protection in Semi-Arid Africa

Petra Tschakert and L. Jen Shaffer

Abstract Resilience is much more than bouncing back after a shock. It also involves the ability of individuals, communities, and entire regions to self-organize and increase their capacity for learning, experimentation, and adaptation. In the context of climate change, a resilience perspective emphasizes learning from the past (memory), monitoring the present, and the ability to anticipate and prepare for the worst. It includes learning to live with change and uncertainty by combining different types of knowledge, envisioning possible futures, and enhancing flexibility in decision-making and planning. Rather than learning by shock, a resilience lens offers a potentially empowering arena for nurturing innovation and the capacity to transform in order to navigate both slow and incremental environmental changes and rapid-onset crises.

This chapter explores the role and potential limits of iterative learning processes for climate change adaptation in rural African communities characterized by high and chronic poverty, coupled with low awareness for complex drivers of change. It stresses learning, memory, creativity, and the need to move forward in spite of imperfect knowledge and vast uncertainties. At the same time, the chapter identifies critical institutional, policy, and power barriers, and potential limits at multiple scales that inhibit just and timely adaptation among vulnerable and marginalized populations, especially those dependent on rainfed agriculture. We identify poverty traps as complex thresholds typified by shifts and losses of key household assets, increasing failure of livelihood response strategies to social and ecological stresses and shocks, ineffective social networks, and limited anticipatory capacity to

P. Tschakert (✉)
Department of Geography, Pennsylvania State University, 322 Walker Building, University Park, PA 16802, USA

Earth and Environmental Systems Institute, Pennsylvania State University, 322 Walker Building, University Park, PA 16802, USA
e-mail: petra@psu.edu

L.J. Shaffer
Department of Anthropology, University of Maryland, College Park, MD, USA

S. Sakai and C. Umetsu (eds.), *Social–Ecological Systems in Transition*, Global Environmental Studies, DOI 10.1007/978-4-431-54910-9_8, © Springer Japan 2014

embrace change, uncertainty, and surprises. We conclude by proposing adaptive social protection as a prospective yet potentially insufficient means for bypassing or escaping poverty traps in the semi-arid tropics of Africa, and facilitating transitions towards livelihood resilience.

Keywords Adaptive social protection • Anticipatory learning • Limits to adaptation • Poverty traps

8.1 Introduction

Resilience is much more than bouncing back after a shock. It also involves the ability of individuals, communities, entire regions, and social–ecological systems, to self-organize and increase their capacity for learning, experimentation, and adaptation. Enhanced creativity, innovation, and the willingness to share and nurture connectedness are considered additional essential ingredients. From a coupled systems perspective, a resilience lens stresses the dynamic interplay of disturbance and reorganization, cross-scalar interactions, and integrated system feedback (Folke 2006). Chapin et al. (2006), for instance, distinguish slow variables (e.g., soil resources, cultural ties to the land) and fast variables (e.g., fire events, population density) that characterize complex social–ecological systems, as well as the institutional responses to these different variables, embedded in social and ecological processes defined by exogenous controls. Understanding the interplay of endogenous and exogenous dynamics and responses is at the core of climate change adaptation and livelihood resilience under climate uncertainty.

In the context of an interdisciplinary project entitled Anticipatory Learning for Climate Change Adaptation and Resilience (ALCCAR), funded by the National Science Foundation,[1] we have been using a resilience perspective to emphasize learning from the past (memory), monitoring the present, and the ability to anticipate and prepare for the worst (Fig. 8.1). We explicitly include learning to live with change and uncertainty by combining different types of knowledge, envisioning possible futures, and enhancing flexibility in decision-making and planning (Tschakert and Dietrich 2010). Rather than learning by shock, we see a resilience lens offering a potentially empowering arena for nurturing innovation and the capacity to transform in order to navigate both slow and incremental environmental changes and rapid-onset crises.

In this chapter we draw upon the ALCCAR project first to illustrate characteristics of livelihood resilience to climate variability and change among subsistence farmers and fisherfolk in Ghana and Tanzania. Then we explore the existence of barriers and limits during the process of adaptation and identify possibly poverty traps likely to hinder certain groups or populations to adapt or transform successfully. Next, we examine two case studies—one from South Africa and the other

[1] NSF-DRU#0826941.

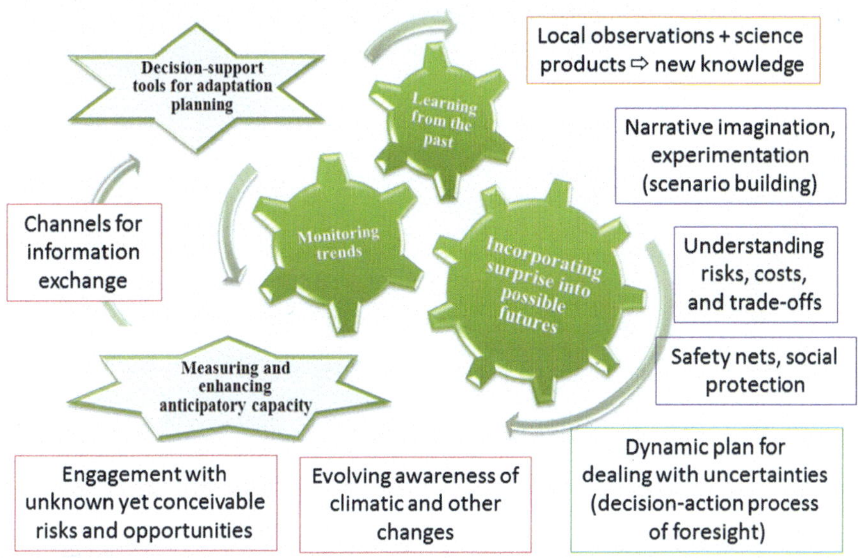

Fig. 8.1 Conceptual framework of anticipatory learning for climate change resilience, designed for the ALCCAR project (after Tschakert and Dietrich 2010)

from Nigeria—to shed light on critical factors which may determine thresholds in livelihood trajectories and implications for avoiding or escaping poverty traps. Finally, we explore the role of adaptive social protection to prevent destitution and collapse, as well as the ethical subtext of dismissing responsibilities.

8.2 Anticipatory Learning and Livelihood Trajectories

The ALCCAR project demonstrates the value, and limitations, of iterative learning processes to enhance adaptive capacity among vulnerable and often marginalized livelihoods in semi-arid regions of Africa. Four years of collaborative work with rural communities in Ghana (Odumase, Xedzoedzoekope, Akeymfour, and Bowiri) and Tanzania (Mlingotini, Makurunge, Chekereni, and Rau) reveal high and chronic poverty coupled with incomplete awareness for complex drivers of change, albeit to different degrees. Through a series of individual and group learning activities, community members, agricultural extension agents, district-level policy makers, and researchers engaged in collective learning. They have drawn upon their memory and creativity, and explicitly acknowledged the need to move forward in spite of imperfect knowledge and vast uncertainties, particularly with respect to climatic and economic futures and the interaction of endogenous and exogenous dynamics that shape rural realities. These activities included constructing historical matrices of past climatic events, scoring of community performance regarding

Fig. 8.2 Participatory mapping showing pace, origin, and control over five key drivers of change, as seen by a women's group in Tanzania (ALCCAR field work)

anticipatory capacity, walking journey landscape interviews, assessments of past and future drivers of change, environmental monitoring, and participatory scenario building.

Following Enfors et al. (2008), who employed future thinking for climate change resilience in Tanzania by building on prior understandings of waves of historic disturbances and shocks, ALCCAR tapped into community memory of past extreme climate events to validate individual and collective experiences, responses, and their varying degrees of success. At least ten droughts and more than a dozen flood events characterize the last 60 years, including extreme events such as the 1984/1985 drought in Tanzania and the 1995 excessive rains in Ghana both of which destroyed fields and crops, triggered human and animal diseases and crop pests, and caused migration, hunger, and death. While these examples showed a suite of response strategies to mitigate harm, their effectiveness varied depending on household assets and the severity of the crisis, which underscores differential adaptive capacity among potentially vulnerable groups.

Shifting from past experiences to anticipate future trajectories revealed deforestation, irregular rainfall, declines in soil fertility, and improved infrastructure and water supply as major changes expected in Ghana. In Tanzania, land scarcity and reduced and unpredictable rains worried most research partners. To understand better the social, ecological, economic, and political drivers that are likely to impact household and community resilience in these communities, we looked specifically at pace, origin, and control over dynamics which will likely drive these changes (Fig. 8.2). Results indicate that drivers expected to occur rapidly or abruptly with their origin outside the community and those beyond people's control were likely to be the most difficult to manage, and hence may represent the most severe threats to community resilience. For Ghana these included cattle encroachment, excessive consumption of alcohol, and the "get-rich-quick syndrome" (exploitive resource use, e.g., for charcoal production). In Tanzania, rising poverty, crop diseases, and unpredictable rains emerged as posing most concerns for livelihood resilience.

Finally, lessons learned through the participatory scenario building exercises suggest that local resource managers and policy makers alike struggle to make sense of the complexities and uncertainties of climate change. We see clear limitations in people's capacity to grasp local climate futures (we used downscaled global climate model projections) and global processes, despite various efforts to introduce, translate, and unpack external science information in various learning cycles. It remains to be tested whether such understandings or lack thereof constitute irrelevant obstacles in the adaptation process or require more and more concentrated efforts to be overcome. In Ghana, distinctly more so than in Tanzania, community members had a firm tendency to create over-idealistic futures with unabated development and strong community unity; it was felt that only exogenous threats could undermine their future.

Similar to Ravera et al. (2011), who use conceptual modeling and participatory scenario development in the context of agropastoralists in semi-arid Nicaragua, our results demonstrate that these tools are vital ingredients for iterative learning cycles and can empower local stakeholders by illustrating opportunities and threats associated with several plausible futures. Moreover, through purposeful envisioning, deliberation, and negotiations over likely trade-offs, they can overcome potential denial, helplessness, hopelessness, and paralyzing fatalism at the backdrop of concurrent challenges. Ravera et al. (2011) further proposed a heuristic analysis of vulnerability and resilience trajectories that visualizes the multiple facets of change in complex regional social–ecological systems (Fig. 8.3). While it appears highly relevant for adaptation planning and policy decision-making, the authors suggest that it requires further refinement for identifying critical thresholds and potential irreversibility.

In this respect, a study by Sallu et al. (2010) on livelihood trajectories and resilience in rural Botswana advances our understanding of the dynamics that households undergo, shifting in and out of vulnerability and quasi-resilient states, depending on their ability to diversify and accumulate livelihood assets. By tracking agro-ecosystem states, access to physical and financial assets, and response capacity over three decades, the authors draw attention to the large majority of dependent households who seem to have no choice other than to follow a degenerative trajectory leading to increased livelihood vulnerability (see Box 8.1). It is through analyses like this that we gain a better understanding of critical thresholds in individual and collective abilities to withstand multiple stressors, including climate change.

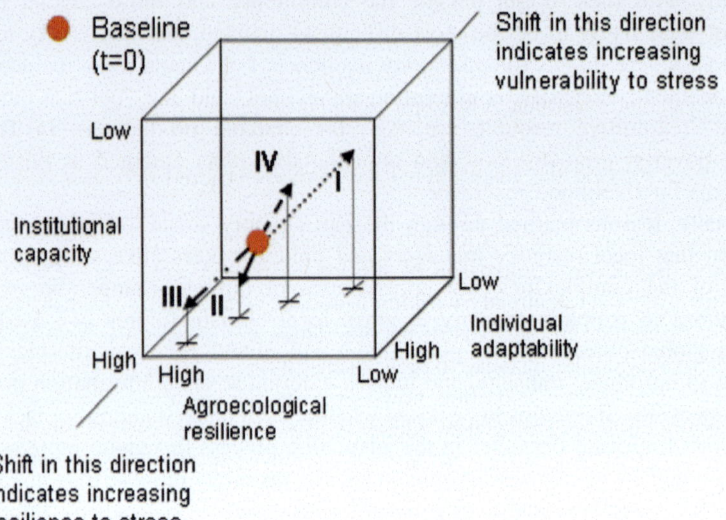

Fig. 8.3 Resilience/vulnerability trajectories (Ravera et al. 2011)

Box 8.1: Example of a Degenerative Livelihood Trajectory in Rural Botswana

T = time; V = vulnerability; R = resilience

T1, 1974: high capacity for agroecosystem to remain productive, high levels of access to natural assets, moderate accumulation of financial and physical assets (livestock), moderate capacity to respond (asset shock)

R1: Engagement in diverse livelihood activities

T2, 1980s: low capacity of agroecosystem to remain productive, retained access to some natural assets (hunting permit), limited access to other natural assets (drought), some physical and financial asset stores, moderate capacity to respond (RAD program support)

V1: Loss of livestock

T3, 1990s: moderate capacity of agroecosystem to remain productive, loss of financial and physical asset stores, loss of access to natural assets (permit changes), increasing reliance on government support

V2: Loss of livelihood activity

T4, 2000a: some capacity of agroecosystem to remain productive, no asset stocks, total reliance on government support

V3: Sole reliance on government support

After Sallu et al. (2010)

8.3 Barriers, Limits, and Traps

Increasingly, adaptation to climate change is understood as inevitable yet not equally accessible, available, and feasible. Lessons from our ALCCAR project indicate limited access to reliable climate information among rural communities and governmental agents, particularly in Ghana, restricted options for livelihood diversification, and the intersection with other livelihood challenges (e.g., increasing crime, difficulties in acquiring new farmland, continuous bush fires) as major obstacles in coping with climate change. These immediate factors interact with policy, institutional, and power barriers, and potential limits at multiple scales which inhibit just and timely adaptation among vulnerable and marginalized populations, especially those dependent on rainfed agriculture. The Fourth Assessment Report of the Intergovernmental Panel on Climate Change (IPCC) offered useful categories of limits (Box 8.2). Barriers are defined as "obstacles that can be overcome with concerted efforts, creative management, change of thinking, prioritization, and related shifts in resources, land uses, institutions, etc." (Moser and Ekstrom 2010: p. 22027). Limits, on the other hand, are "obstacles that tend to be absolute in a real sense: they constitute thresholds beyond which existing activities, land uses, ecosystems, species, sustenance, or system states can no longer be maintained, not even in a modified fashion. Beyond such limits looms irreversible loss (and the adjustment to living with that loss) and/or radical system shifts, including innovation and novelty" (Moser 2009: p. 33). Adger et al. (2009) argue that many seeming limits are in fact malleable barriers which could be overcome with political will, adequate resources, and social support; this is particularly true for social limits. Jones and Boyd (2011) provide a useful categorization of social barriers, distinguishing between cognitive behavior, normative behavior, and institutional structure and governance. Peterson (2009) further examines ecological limits while O'Brien (2009) considers the role of values in subjectively defining limits to adaptation.

Box 8.2: Different Times of Limit to Adaptation

Physical and ecological limits (thresholds in the resilience of kelp forest ecosystems, coral reefs, rangelands and lakes affected both by climate change and other pollutants; rapid sea-level rise and transformation of islands; droughts in sub-Saharan Africa leading to land degradation, diminished livelihood opportunities, food insecurity, internal displacement of people, cross-border migrations, and civil strife; loss of key stone species; regime shifts in ecosystems)

Technological limits (technologically possible vs economically feasible and culturally desirable; not accessible to all—increased inequalities and side-effects for others)

(continued)

Box 8.2 (continued)

Financial barriers (local poverty; enormous international costs for climate-proofing; even barriers to climate-risk insurance)

Information and cognitive barriers (more knowledge doesn't mean action—attitude-behavior gap; differential risk perceptions and priorities for subjective, immediate, and known risks; divergence between perceived and real adaptive capacity; public confusions, appealing to fear and guilt)

Social and cultural barriers (different risk tolerances, different preferences about measures depending on worldviews, values, and beliefs; differential power and access to decision-making)

After Adger et al. (2007) (IPCC, WG II, Chap. 17)

Exploring the multiple angles of adaptation has offered a timely lens to scrutinizing what so far has been largely defined from a techno-economic perspective—infrastructural interventions, technological innovations, and cost-benefit analyses. To zoom in further on key processes, Thornton and Manasfi (2010) propose a systematic assessment of actions that people typically undertake during adaptation (mobility, exchange, rationing, pooling, diversification, intensification, innovation, and revitalization), and factors blocking these actions. Moser and Ekstrom (2010), in a somewhat different approach, propose a framework that identifies barriers at various stages of three distinct phases of the adaptation process—understanding, planning, and managing. These include thresholds of concern for detecting a problem, level of agreement to (re-)define the problem, authorization to implement options, ability to monitor outcomes, leadership to develop options, and thresholds of concern over possible negative consequences.

Drawing attention to a multitude of barriers and potential limits, especially those related to social, cultural, and institutional dimensions, introduces both a long overdue social framing of adaptation and a humanizing lens exposing pre-existing inequalities and injustices, chronic poverty, disempowerment, and structural violence, all of which hamper successful adaptation. Perhaps more importantly still, an explicit focus on barriers and limits allows connecting the adaptation community with both the development and the resilience community, both of whom have been examining poverty and poverty traps, even though from a slightly different angle. We argue that linking these conceptual worlds more explicitly provides useful insights into social, ecological, and potentially moral thresholds and their interconnections in complex and coupled systems. More specifically, we propose to identify critical thresholds in livelihood trajectories, expanding on Sallu et al. (2010), which indicate desirable pathways ("upwardly mobile") and undesirable pathways (potential descent into poverty traps), and prospective interventions (e.g., adaptive social protection) to counteract what otherwise may appear as a deterministic course (Fig. 8.4). We describe the concept of poverty traps, then provide two examples from Africa on how such traps can be avoided, and finally assess how social protection could assist in this effort.

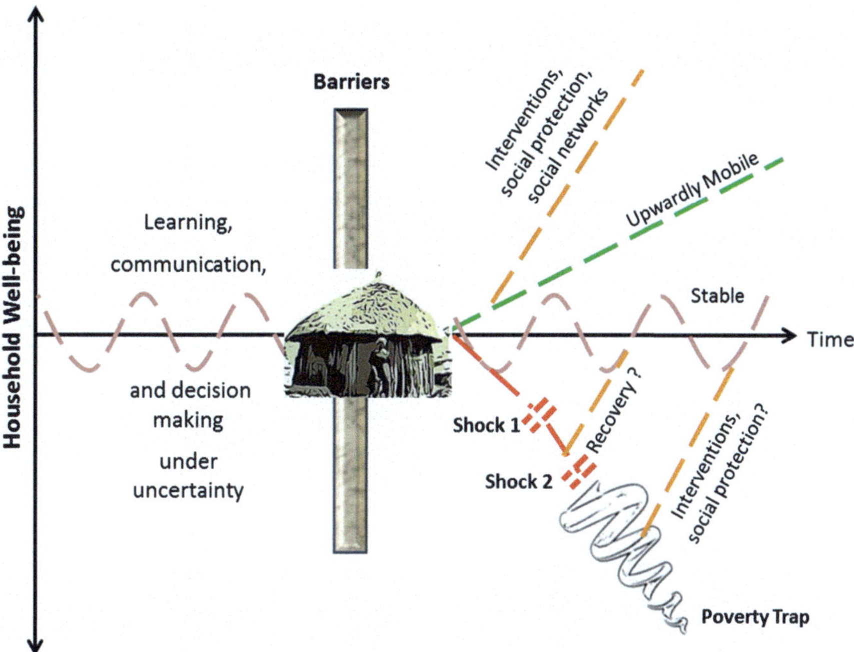

Fig. 8.4 Schematic illustration of possible livelihood trajectories, showing barriers and thresholds which could lead to upward mobility or a downward spiral into poverty traps through a series of shocks

From the perspective of development economics, poverty traps are typically understood as situations in which people are impoverished through a self-reinforcing mechanism and with few if any chances to break the vicious cycle; more specifically, they refer to situations in which individuals and households holding highly unproductive asset portfolios trap themselves in chronic poverty, despite rational attempts to manage risk (Barrett and McPeak 2006).

In the context of resilience thinking and dynamic social–ecological systems, often characterized through the concept of adaptive cycles (Gunderson and Holling 2002), a poverty trap represents a situation of persistent maladaptation, or one type of pathological state of the adaptive cycle (Allison and Hobbs 2004). It is characterized by low connectedness and potential and, despite abundant and promising ideas, leadership to channel these ideas is absent and the possibility for change is not realized (Fig. 8.5). Carpenter and Brock (2008) further describe poverty traps as constellations with high heterogeneity of entities, high capacity to explore, yet low capacity to focus and high capacity to dissipate stress. So, seen from a resilience perspective, poverty traps are not confined to states of economic deprivation; they can apply to dysfunctional institutional settings, social–ecological systems in situations of chronic or recurrent disaster, or systems which undergo huge fluctuations but scatter stress before adaptive action can occur. In other words, poverty traps are

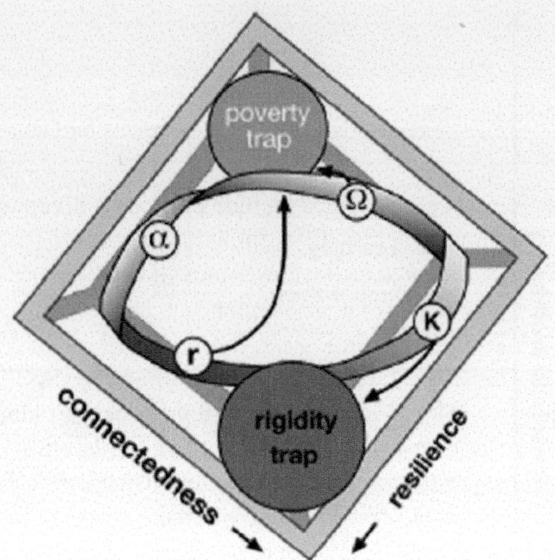

Fig. 8.5 Poverty traps as one type of pathological state of the adaptive cycle (Allison and Hobbs 2004)

unrealized potential (Carpenter and Brock 2008), a condition inhibiting any type of transformation, a form of being stuck (trapped) despite initial possibilities.

We build on these two notions of poverty traps to examine first household transitions into such traps and then consider possible pathways out, through the use of two case studies from Africa. The purpose is to identify critical elements and processes shaping such trajectories. In turn, they could provide additional evidence for critical thresholds in the context of livelihood and social-ecological system resilience under climatic uncertainty and change. We will end by exploring the role of support structures such as social networks and social protection for avoiding and/or escaping poverty traps.

8.4 Avoiding Poverty Traps in Rural Africa

Our general argument is that, below a certain threshold of assets, households fall into poverty traps because they cannot accumulate enough to maintain their well-being and practice livelihood activities successfully. This threshold of assets varies spatially and temporally, but essentially supports household response strategies to stresses and shocks, and the adaptive capacity to embrace change, uncertainty, and surprise. The kind of human, social, natural, physical, and financial capital available to households is just as important as the amount available and the capability with which households can flexibly juggle their various assets to generate a stream of benefits (Bebbington 1999; Barrett et al. 2005). In the following case study summaries, we assess the importance of assets and strategies in avoiding poverty traps and propose ways to safeguard household transitions into such traps.

8.4.1 Post-Apartheid South Africa

Many South Africans thought economic disparities between whites and blacks would end with Apartheid. However, widespread poverty, minimal access to basic services, and huge inequalities in resource distribution greeted the ANC's 1994 rise to power (Adato et al. 2004; Casale and Desmond 2007). Household inequality and poverty continued to worsen over the following decade, not improve. From 1993 to 2005, unemployment rose as wages fell, the number of households receiving social grants for child and elder support rose, remittances dropped, and although access to public service assets like electricity, sanitation, water, and formal housing improved, the quality was low (Bhorat and Kanbur 2006; Casale and Desmond 2007). Rural and black South African households suffered most. The legacy of an Apartheid economy "in which social exclusion and poverty continue to interact in a mutually self-sustaining fashion" remained (Adato et al. 2004: p. 2).

Households often cope with stresses and shocks by drawing on bonding capital and social networks generally composed of kin and close friends to access remittances, labor assistance, and tools, save and borrow cash, bury family, maintain food security, and receive emotional support (Putnam 2000). Social networks with widespread connections outside kin and close friends, having more bridging capital, offer information about available jobs, employer contacts, and job seeking advice, as well as accommodation and transportation during interviews. Various poverty analyses found support for the legacy hypothesis. Households trapped in structural poverty lack the effective social network connections associated with bridging capital that provide the information and connections needed to find good employment and financial opportunities (Adato et al. 2004). These marginalized households often had no access to wages or remittances, or could only find casual labor (Carter and May 1999). Downwardly mobile households lost financial assets via catastrophe, failed investments, reneged remittances, or death of a wage earner, and could not recover because they had nothing on which to build and lacked the necessary social connections to obtain more. However, those households with access to public service assets and government financial support like old age pensions or child support could stabilize and potentially move out of poverty traps (Bhorat and Kanbur 2006; Carter and May 1999). Stable, non-poor households leveraged greater social bridging capital and financial capital by working at multiple formal, informal, and casual employment opportunities. The upwardly mobile households studied invested in education for future payoffs and frequently started businesses with funds gained when their former employers closed shop (Adato et al. 2004; Carter and May 1999).

Multiple, interacting stressors and shocks limited both the short- and long-term ability of households to stabilize themselves and move out of poverty by reinforcing the social and economic dimensions of poverty traps. The HIV/AIDS crisis removed active skilled and semi-skilled adults from the labor pool, leaving impoverished households of orphaned children and elders. As families sought to meet short-term survival needs, the epidemic further undermined long-term resilience affecting property transfer, children's access to education, and family networks (Drimie and Casale 2008). Simultaneously, feedback from rising criminal activity contributed to further income inequality as investor uncertainty limited job creation and economic growth (Bhorat and Kanbur 2006). Climate change and governance issues also increased household vulnerabilities while undermining adaptive capacity (Drimie and Casale 2008). The barriers and obstacles surrounding post-Apartheid poverty traps were very strong, but not insurmountable for households with connections to effective social networks, access to formal employment, and small amounts of financial capital.

8.4.2 Kofyar Households in Post-independence Nigeria

In 1953, Kofyar farmers began migrating south from their homeland settlements on the Jos Plateau to settle 40 km away on a relatively vacant, forested frontier. Kofyar households flourished on this frontier despite the domestic food insecurity, spiraling imports, inflation, rapid population growth, and little or no government support for the lower economic classes that many of their fellow Nigerians suffered from throughout the 1960s and 1970s (Netting et al. 1989). By the mid-1980s, home farms on the Jos Plateau became depopulated outposts anchored by elders, schools, vacation homes, and traditional ceremonies, as Kofyar outmigration accelerated after Nigerian independence (Stone 1998). Revitalization of home farm settlements on the plateau in the mid-1990s stemmed from a desire to maintain ethnic identity and access government resources. In general, Kofyar household well-being has remained stable or prospered over time.

Several factors helped Kofyar farming households succeed amidst a growing nation's stresses and shocks. Acquiring new lands on the frontier cost little to nothing, and a ready domestic market for agricultural production existed. Good access to markets, time to migrate and cultivate with no outside pressures, and safe incentives to make money, like education, medical treatment, and desirable manufactured goods, also encouraged migration (Netting et al. 1989; Stone 1998). A distinct lack of national government interference ensured that households relied on their indigenous agro-system knowledge, cultural institutions, and highly-bonded social networks as they moved out onto the frontier (Netting et al. 1989). Households on the Jos Plateau traditionally employ intensive, non-gendered cultivation techniques, hand tools, deep agro-ecosystems knowledge, and organized labor to farm small plots. Emigrants transferred these strategies to the frontier's extensive agricultural areas to grow traditional cultigens, as well as cash crops of rice and yams that complemented home farm production in the uplands. Indigenous knowledge, gained while farming the Jos Plateau, drove innovations including two planting seasons per year, inter-cropping, and trying new varietals. Polygynous practices and multiple family formations, distinct from the nuclear families of the home farms, aided Kofyar household sizes to expand to fulfill labor needs (Netting et al. 1989). Traditional labor mobilization strategies of beer parties and reciprocal labor groups also helped households meet the year round demands of growing

Table 8.1 Comparison of key elements in two poverty trap constellations

	Post-Apartheid South Africa	Kofyar, post-independence Nigeria
Shocks and Stresses	• HIV/AIDS epidemic • Rising criminal activity • Governance issues • Climate change • Legacy of social exclusion and poverty	• Domestic food insecurity • Spiraling imports • Inflation • Rapid population growth • Lack of government support
Assets	• Public services (electricity, water, sanitation, formal housing) • Old age pensions, child support, etc. • Payoffs from former employer • Social networks	• Indigenous knowledge • Cultural labor institutions • Free or low cost land available • Ready national market for products • Time (migrate, develop fields, etc.) • Social networks
Strategies	• Portfolio of formal, informal, and casual employment • Start small businesses • Expand and strengthen social networks • Invest in children's education • Depend upon remittances	• Traditional labor mobilization • Expansion of household size • Acquire new fields slowly, as desired • Non-gendered cultivation work • Plant twice a year, try new varietals, intensive cropping, intercropping
Social networks	*Effective* networks include non-kin, urban, and employment related contacts	*Effective* networks include kin and close friends for labor mobilization
Transition	Mixed Depends on access to effective social networks and formal employment	Maintain and/or improve household well-being

multiple crops for personal and market use (Stone et al. 1990). Economic inequality remained low in Kofyar communities as upwardly mobile households could move onto plentiful, open land and grow more cash crops to increase their income and labor force (Stone et al. 1984).

8.4.3 Lessons Learned

Differences between the case studies offer a view into what helps households move out of poverty, remain stable, and even prosper in the midst of social and ecological stresses and shocks (Table 8.1). Effective social networks, with connections bridging outside the immediate family and close friends, helped households leverage the financial assets necessary for finding good, formal employment, as well as additional social connections in South Africa. Kofyar farmers used bonding capital in social networks of family and neighbors to organize the labor necessary for expanding market production and increasing household income. Institutional social protection and safety nets also play a role in bypassing or escaping poverty traps. Impoverished South African households with access to government financial safety nets and public service assets avoided traps and, in some cases, stabilized enough to embark on a path of upward mobility. In the case of Kofyar, a complete lack of interference by the Nigerian government required people to pull together as a community and draw upon cultural knowledge and institutions to prosper.

It would seem that our lessons learned are at odds. However, the Kofyar case study describes the experience of a single cultural group living within a small region while the South African case study deals with household poverty among more culturally diverse peoples spread across a larger national landscape. Development of effective adaptive social protection measures must consider social, temporal, and spatial scales in order to prevent people falling into poverty traps and facilitate transitions towards livelihood resilience. We explore such measures in the final section.

8.5 Adaptive Social Protection

We have examined poverty traps as complex thresholds typified by shifts and losses of key household assets, ineffective social networks, and limited anticipatory capacity to embrace change, uncertainty, and surprises. An additional factor, although not as apparent in the case studies, is increasing failure of livelihood response strategies to social and ecological stresses and shocks. Drawing upon the Botswana case study conducted by Sallu et al. (2010), we argue that such "slowing down of response capacity" is similar to the slowing down described by Dakos et al. (2008: p. 14311) as a "universal property of systems approaching a tipping point," including potentially those in social systems. This analogy, although not tested for household level response options per se, mirrors Tainter's (1988) depiction of poverty traps in the context of collapsing complex societies as situations in which sources of novelty are gradually diminished, leading to self-eroding capacity for adaptive responses. We find this an intriguing area for future research. In this final section, we explore the role of adaptive social protection as a prospective means for bypassing or escaping poverty traps. We apply emerging insight to facilitate the conceptualization of transitions towards livelihood resilience in the semi-arid tropics of Africa.

Standard social protection (SP) describes all public and private initiatives which provide income or consumption transfers to the poor, protect the vulnerable against livelihood risks, and enhance the social status and rights of the marginalized (Devereux and Sabates-Wheeler 2004). The key goal is to extend the benefits of economic growth to poor, vulnerable, and marginalized groups, or at least reduce their economic and social vulnerability. SP measures are typically presented in four different categories: provision, prevention, and protection of assets, as well as transformation of social relations and rights (Table 8.2).

The concept of adaptive social protection (ASP) goes further; it provides a framework for more effective integration of standard SP, climate change adaptation, and disaster risk reduction into *one* coherent approach (Davies et al. 2009a, b).

Table 8.2 Types of traditional social protection programs (SP) and instruments, as well as potential and simultaneous benefits for climate change adaptation and disaster risk reduction (after Davies et al. 2009b)

SP category	SP instruments	Adaptation and DRR benefits
Protective (coping strategies)	• Social service provision • Social transfers (food/cash), including safety nets • Social pension schemes • Public works programs	➢ Protection of those most vulnerable to climate risks, with low levels of adaptive capacity
Preventive (coping strategies)	• Social transfers • Livelihood diversification • Weather-indexed crop insurance • Social insurance	➢ Prevention of damaging cropping strategies as a result of risks to weather-dependent livelihoods
Promotive (building adaptive capacity)	• Social transfers • Access to credit • Asset transfers or protection • Starter packs (drought/flood-resistant) • Access to common property resources. • Public works programs	➢ Promotion of resilience through livelihood diversification and security to withstand climate-related shocks ➢ Promotion of opportunities arising from climate change
Transformative (building adaptive capacity)	• Promotion of minority rights • Anti-discrimination campaigns • Social funds • Proactively challenging discriminatory behaviour	➢ Transformation of social relations to combat discrimination underlying social and political vulnerability

Although SP is perceived as a vital ingredient for the delivery of pro-poor climate change adaptation and disaster risk reduction among vulnerable groups, mainly in developing countries (Heltberg et al. 2009; Stern 2007), SP programs have traditionally focused on the chronic poor, with potential side benefits for climate-related stresses. Under ASP, such programs would become more flexible and dynamic to capture better both transient and chronic poor affected by increased climate variability, longer-term climate risks, extreme climatic events, and other disasters. This is particularly relevant as neo-liberal policies have progressively eroded social safety nets and other state-led support structures. However, concrete examples of successful implementation of truly integrated ASP programs are so far rather sparse. However, lessons can be learned from programs such as Ethiopia's nation-wide Productive Safety Net Programme which helps chronically impoverished individuals. It has shown protection of existing household assets, a decline in "distress selling" of assets, and positive effects on household food consumption (Slater et al. 2006; Devereux et al. 2006; Ellis et al. 2009), all of which also strengthen adaptive capacity under climatic calamities.

We argue that classic safety net programs and more recent adaptive social protection programs can fulfill a double purpose (see also Fig. 8.4): one is to assist households as a point of entrance to an "upwardly mobile" trajectory that takes advantage of an emerging window of opportunity at a critical threshold; the other, perhaps even more important, is to assist vulnerable households and marginalized groups to find their way out of persistent poverty traps after a series of experienced shocks and back onto a fairly stable trajectory. While market-based adaptation options tend to exclude those who are constantly poor, ASP programs offer tailor-made schemes and interventions to both the chronic and shifting poverty contexts (IDS 2009).

8.6 Conclusion

What lessons have we learned from our ALCCAR project in Ghana and Tanzania, the research by Sallu et al. (2010) in Botswana, and the two case studies on poverty traps in South Africa and Nigeria? We distill the following key messages. (1) There are multiple exogenous and endogenous dynamics and responses which shape the vulnerability and resilience of small-scale farming and fishing communities in semi-arid regions of Africa; some may seem trivial, such as excessive alcohol consumption and limited access to reliable climate information; yet, they weaken adaptive capacity in the face of multiple threats and contribute to the persistence of poverty traps and, hence, should receive enhanced attention. (2) We need a more sophisticated understanding of differential thresholds in diverse livelihood trajectories, by households, stakeholder groups, and entire agro-ecological regions, those that lead to upward mobility and others that shift individuals and communities into poverty traps. (3) Social networks and adaptive social protection programs are vital to facilitate upward mobility and allow the less fortunate to escape poverty traps. (4) In spite of conceptual advances with respect to adaptive social protection, political will and adequate resources are lacking to implement far-reaching programs to assist vulnerable groups in most if not all of semi-arid Africa, and beyond. (5) Global market mechanisms and social support structures at local scales alone will not be sufficient to facilitate transitions towards livelihood resilience. What is urgently needed is a holistic understanding of multi-faceted vulnerability and deprivation, as well as unrealized adaptive potential and, subsequently, an ethical commitment and the responsibility to protect all those who already face the limits to (autonomous) adaptation.

Acknowledgements We would like to thank the members of the ALCCAR project team (NSF-DRU award # 0826941) and participants in the June 2011 Worldwide Universities Network (WUN) workshop on "Limits to Adaptation" for extensive discussions which stimulated our thinking about livelihood trajectories, thresholds, poverty traps, and social protection.

References

Adato M, Carter M, May J (2004) Sense in sociability? Social exclusion and persistent poverty in South Africa. Staff Paper Series 477, Agricultural and Applied Economics, University of Wisconsin. http://ideas.repec.org/p/ecl/wisagr/477.html

Adger WN, Agrawala S, Mirza MMQ, Conde C, O'Brien K, Pulhin J, Pulwarty R, Smit B, Takahashi K (2007) In: Parry ML, Canziani OF, Palutikof JP, van der Linden PJ, Hanson

CE (eds) Assessment of adaptation practices, options, constraints and capacity. Climate change 2007: impacts, adaptation and vulnerability. Contribution of working group II to the fourth assessment report of the intergovernmental panel on climate change. Cambridge University Press, Cambridge, pp 717–743

Adger WN, Dessai S, Goulden M, Hulme M, Lorenzoni I, Nelson DR, Naess LO, Wolf J, Wreford A (2009) Are there social limits to adaptation to climate change? Clim Change 93:335–354

Allison HE, Hobbs RJ (2004) Resilience, adaptive capacity, and the "lock-in trap" of the Western Australian agricultural region. Ecol Soc 9(1):3. http://www.ecologyandsociety.org/vol9/iss1/art3/

Barrett C, McPeak J (2006) Poverty traps and safety nets. Poverty Inequal Dev 1:131–154

Barrett C, Bezuneh M, Clay D, Reardon T (2005) Heterogeneous constraints, incentives and income diversification strategies in rural Africa. Quart J Int Agric 44(1):37–60

Bebbington A (1999) Capitals and capabilities: a framework for analysing peasant viability, rural livelihoods and poverty. World Dev 27(12):2021–2044

Bhorat H, Kanbur R (2006) Poverty and well-being in post-Apartheid in South Africa: an overview of data, outcomes and policies. In: Bhorat H, Kanbur R (eds) Poverty and policy in post-Apartheid South Africa. HSRC Press, Pretoria, pp 1–18

Carpenter SR, Brock WA (2008) Adaptive capacity and traps. Ecol Soc 13(2):40. http://www.ecologyandsociety.org/vol13/iss2/art40/

Carter M, May J (1999) One kind of freedom: poverty dynamics in post-Apartheid South Africa. Staff Paper Series 427, Agricultural and Applied Economics, University of Wisconsin. http://www.aae.wisc.edu/pubs/sps/pdf/stpap427.pdf

Casale D, Desmond C (2007) The economic well-being of family: households' access to resources in South Africa. In: Amoateng A, Heaton T (eds) Families and households in post-Apartheid South Africa, 1995–2003: socio-demographic perspectives. HSRC Press, Cape Town, pp 61–88

Chapin FS III, Lovecraft AL, Zavaleta ES, Nelson J, Robards MD, Kofinas GP, Trainor SF, Peterson GD, Huntington HP, Naylor RL (2006) Policy strategies to address sustainability of Alaskan boreal forests in response to a directionally changing climate. Proc Natl Acad Sci U S A 103(45):16637–16643

Dakos V, Scheffer M, van Nes EH, Brovkin V, Petroukhov V, Held H (2008) Slowing down as an early warning signal for abrupt climate change. Proc Natl Acad Sci U S A 105(38):14308–14312

Davies M, Guenther B, Leavy J, Mitchell T, Tanner T (2009a) Climate change adaptation, disaster risk reduction and social protection: complementary roles in agriculture and rural growth? IDS Working Paper 320, IDS, Brighton

Davies M, Oswald K, Mitchell T (2009b) Climate change adaptation, disaster risk reduction, and social protection. In: Promoting pro-poor growth: social protection. OECD, pp 201–217. http://www.oecd.org/dataoecd/63/10/43514563.pdf

Devereux S, Sabates-Wheeler R (2004) Transformative social protection, IDS Working Paper 232, IDS, Brighton

Devereux S, Sabates-Wheeler R, Tefera M, Taye H (2006) Ethiopia's productive safety net programme: trends in PSNP transfers within targeted households, IDS Working Paper 232, IDS, Brighton

Drimie S, Casale M (2008) Families' efforts to secure the future of their children in the context of multiple stresses including HIV and AIDS. Report for the Joint Learning Initiative on Children and HIV/AIDS, p 83

Ellis F, Devereux S, White P (2009) Social protection in Africa. Edward Elgar, Cheltenham

Enfors EI, Gordon LJ, Peterson GD, Bossio D (2008) Making investments in dryland development work: participatory scenario planning in the Makanya catchment, Tanzania. Ecol Soc 13(2):42. http://www.ecologyandsociety.org/vol13/iss2/art42/

Folke C (2006) Resilience: the emergence of a perspective for social-ecological systems analyses. Glob Environ Chang 16:253–267

Gunderson LH, Holling CS (2002) Panarchy: understanding transformations in human and natural systems. Island Press, Washington, DC
Heltberg R, Siegel PB, Jorgensen SL (2009) Addressing human vulnerability to climate change: toward a "no-regrets" approach. Glob Environ Chang 19(1):89–99
IDS (2009) Linking climate change adaptation, disaster risk reduction and social protection. CBA workshop, Institute for Development Studies, Brighton, UK
Jones L, Boyd E (2011) Exploring social barriers to adaptation: insights from Western Nepal. Glob Environ Chang 21(4):1262–1274
Moser S (2009) Governance and the art of overcoming barriers to adaptation. IHDP Update 3:31–36
Moser SC, Ekstrom JA (2010) A framework to diagnose barriers to climate change adaptation. Proc Natl Acad Sci U S A 107(51):22026–22031
Netting R, Stone P, Stone G (1989) Kofyar cash-cropping: choice and change in indigenous agricultural development. Hum Ecol 17(3):299–319
O'Brien K (2009) Do values subjectively define the limits to climate change adaptation? In: Adger WN, Lorenzoni DI, O'Brien KL (eds) Adapting to climate change: thresholds, values, governance. Cambridge University Press, Cambridge, pp 164–180
Peterson G (2009) Ecological limits to adaptation to climate change. In: Adger WN, Lorenzoni DI, O'Brien KL (eds) Adapting to climate change: thresholds, values, governance. Cambridge University Press, Cambridge, pp 25–41
Putnam R (2000) Bowling alone: the collapse and revival of American community. Simon & Schuster, New York
Ravera F, Tarrasón D, Simelton E (2011) Envisioning adaptive strategies to change: participatory scenarios for agropastoral semiarid systems in Nicaragua. Ecol Soc 16(1):20. http://www.ecologyandsociety.org/vol16/iss1/art20/
Sallu S, Twyman C, Stringer L (2010) Resilient or vulnerable livelihoods? Assessing livelihood dynamics and trajectories in rural Botswana. Ecol Soc 15(4):3. http://www.ecologyandsociety.org/vol15/iss4/art3/
Slater R, Ashley S, Tefera M, Buta M, Esubalew D (2006) Ethiopia productive safety net programme (PSNP): PSNP policy. Programme and Institutional Linkages, Overseas Development Institute, London
Stern N (2007) The economics of climate change: the Stern review. Cambridge University Press, Cambridge
Stone G (1998) Keeping the home fires burning: the changed nature of householding in the Kofyar homeland. Hum Ecol 26(2):239–265
Stone G, Stone P, Netting R (1984) Household variability and inequality in Kofyar subsistence and cash-cropping economies. J Anthropol Res 40(1):90–108
Stone G, Netting R, Stone P (1990) Seasonality, labor scheduling, and agricultural intensification in the Nigerian savanna. Am Anthropol 92(1):7–23
Tainter JA (1988) The collapse of complex societies. Cambridge University Press, Oxford
Thornton TF, Manasfi N (2010) Adaptation – genuine and spurious: demystifying adaptation processes in relation to climate change. Environ Soc 1(10):132–155
Tschakert P, Dietrich K (2010) Anticipatory learning for climate change adaptation and resilience. Ecol Soc 15(2):11. http://www.ecologyandsociety.org/vol15/iss2/art11/

Chapter 9
Dynamics of Social–Ecological Systems: The Case of Farmers' Food Security in the Semi-arid Tropics

Chieko Umetsu, Thamana Lekprichakul, Takeshi Sakurai, Taro Yamauchi, Yudai Ishimoto, and Hidetoshi Miyazaki

Abstract Resilience is defined as "the capacity of a system to experience shocks while retaining essentially the same function, structure, feedbacks, and therefore identity (Walker et al. (2004) Ecol Soc 9(2):5). Although resilience has been defined and analyzed in ecological as well as social-ecological terms, their method of analysis is still under development. Recently, the concept of resilience has been directly applied to regional development and food security issues where people's livelihoods rely heavily on the natural resource base. Resilience of social-ecological system (SES) is considered an important component for achieving sustainability.

Within Semi-Arid Tropical Sub-Saharan Africa, communities' livelihoods depend critically on fragile and poorly endowed natural resources, and poverty and environmental degradation are widespread. People in these regions depend largely on rain-fed agriculture, and their livelihoods are vulnerable to environmental variability. Environmental resources such as vegetation and soil are also vulnerable to human activities. To surmount these environmental challenges, human society and ecosystems must have a capacity to recover quickly from environmental shock.

We argue that, in order to operationalize resilience, it is important for us to consider *resilience* in the context of human security of rural households in semi-arid tropics (SAT) regions. We consider resilience *to* environmental variability, such as

C. Umetsu (✉)
Research Institute for Humanity and Nature, Kyoto, Japan

Nagasaki University, Nagasaki, Japan
e-mail: umetsu@nagasaki-u.ac.jp; umetsu@chikyu.ac.jp

T. Lekprichakul • Y. Ishimoto • H. Miyazaki
Research Institute for Humanity and Nature, Kyoto, Japan

T. Sakurai
Hitotsubashi University, Tokyo, Japan

T. Yamauchi
Hokkaido University, Sapporo, Japan

S. Sakai and C. Umetsu (eds.), *Social–Ecological Systems in Transition*, Global Environmental Studies, DOI 10.1007/978-4-431-54910-9_9, © Springer Japan 2014

drought, flooding, and social changes. We consider resilience *of* food supply and consumption, health status, agricultural production, and livelihoods. Lastly, we consider resilience *for* protecting human security, i.e., survival, livelihoods, and dignity. The purpose of the chapter is to show our empirical evidence from Zambia and the dynamics of farmers' livelihoods in response to various shocks, discuss whether threshold can be defined in the context of food security in social-ecological system, and, lastly, investigate the role of institutions to build adaptive capacity of the communities.

Keywords Adaptive capacity • Agricultural system • Environmental shock • Food security • Resilience

9.1 Introduction

Resilience is defined as "the capacity of a system to experience shocks while retaining essentially the same function, structure, feedbacks, and therefore identity" (Walker et al. 2004, 2006a). Resilience, in other words, refers to the largest amount of disturbance that a system can endure without changing the original steady state and without moving into an alternate regime. The social-ecological system, which is an important system for considering resilience, has a certain threshold. Once the threshold is crossed, leading to cascading changes, some systems are able to respond to disturbances by maintaining a reversible regime shift over time, but other regime shifts are irreversible. A system is said to be more resilient if it has the ability to absorb larger disturbances without moving into an alternate regime.

The concept of ecological resilience has been a focus of ecological research since it was defined in the seminal paper "Resilience and Stability of Ecological Systems" by Holling (1973). The earlier concept of resilience was called *engineering resilience*, where resilience is defined as the recovery time of an ecological system to the pre-disturbance equilibrium condition. The shorter the return time, the greater the resilience of the ecological system. The equilibrium concept was expanded to the concept of *ecological resilience*, which emphasizes the capacity to endure disturbance, incorporating non-linearity, multiple equilibrium, and regime shifts. After the 1990s, the resilience concept focused more on the self-reorganizing properties after the disturbance. Recently, researchers applied the resilience concepts used in ecology and engineering to complex social–ecological systems (Levin et al. 1998; Levin 1999; Berkes and Folke 1998; Berkes et al. 2003; Gunderson 2003; Gunderson et al. 2006). Resilience is an especially relevant concept for considering the recovery of disaster-affected communities and the development of rural societies, where livelihoods are highly dependent on a natural resource base.

The development of the ecological resilience theory occurred in parallel with the emergence of the field of ecological economics, which was established in the late 1980s. Ecological economics arose mainly in the developed world and focused less on critical development issues such as poverty and environmental degradation. In contrast, conventional development economics ignored ecosystem services which form the basis of human economic activity. There was thus a need to link socio-economic research with ecological research and to apply the resilience concept in social–ecological systems to address development issues such as resource degradation and enhance human security. To monitor and manage resilience, it is necessary to understand how the system adapts to different disturbances and system properties in terms of its buffer stock or redundancy, flexibility, margin to its threshold, and system tolerance when approaching the boundary. Although attempts have been made to analyse resilience both as an ecological system and an integrated social-ecological system (Holling 1973; Folke 2006), the method of resilience analysis is still under development. Recently, the concept of resilience has been directly applied to regional development and food security issues in a setting where people's livelihoods rely heavily on the natural resource base (Perrings 2006; Mäler 2008; WRI 2008; ICRISAT 2010; Umetsu 2011). Resilience of the social-ecological system (SES) is considered an important component for achieving sustainability (ICSU 2010).

Within the semi-arid tropics (SAT) of Sub-Saharan Africa, food security, livelihood resilience, and poverty reduction are critical development issues. People's livelihoods critically depend on fragile and poorly endowed natural resources, and poverty and environmental degradation are widespread. Agricultural systems in these regions are largely rain-fed, and people's livelihoods are vulnerable to environmental variability. Environmental resources such as vegetation and soil are also vulnerable to human activities. To surmount these environmental challenges, human society and ecosystems must strengthen their capacity to recover from environmental shocks quickly.

To operationalize resilience research, the concept must be contextualized by clearly defining resilience of what, resilience to what, and resilience for what or for whom in the examination (Walker et al. 2006a, b). We argue that with rising population, depletion of natural resources, degradation of the natural environment, and increased climate variability, human insecurity in general and food insecurity in particular of rural households in the SAT region is a high-priority development challenge to be collectively addressed by local governments and international communities. We consider the resilience *of* food supply and consumption, health status, agricultural production, and livelihoods; resilience *to* environmental variability, such as drought, flood, and social changes; and resilience *for* protecting human security, which has three pillars, i.e., survival, livelihood, and dignity.

The purpose of this chapter is to show our empirical evidence from Zambia and the dynamics of farmers' livelihoods in response to various shocks. First, we describe the social-ecological system in the context of food security in the SAT. Next, we show some examples of the quasi-threshold to food insecurity in the social-ecological system and discuss whether the threshold can be defined. Quasi-threshold is used to mean that crossing the limit does not necessarily move the system to another regime. Then we analyze resilience indicators for assessing general resilience for monitoring and management purposes. Finally, we provide policy implications for enhancing resilience for food security. The role of institutions to build adaptive capacity is also discussed.

9.2 Social–Ecological Systems and Food Security in the SAT

The SAT is located in 48 countries across four continents, covering 22.6 billion km^2 and 15.2 % of land area. In Sub-Saharan Africa, including the southern belt south of the Sahara desert, the SAT region covers most of eastern and south-central Africa. The ecologically oriented definition of SAT was suggested by Troll (1965). He considered the length of the dry season, the length and quality of the wet season, and rainfall adequacy to meet evapotranspiration needs; thus, this definition is oriented to agronomic management. The wet season includes the months when the amount of rainfall exceeds the maximum possible evapotranspiration. According to Troll's definition, the SAT is a tropical region with a wet season of 2–7 months and dry season of 5–10 months. Recently, Ryan and Spencer (2001) defined SAT as follows: (1) length of the growing period is 75–180 days, (2) mean monthly temperature for all months exceeds 18 °C, and (3) daily mean temperature during the growing period is above 20 °C. In the SAT region, the livelihoods of communities critically depend on fragile and poorly endowed natural resources, and poverty and environmental degradation are widespread. People in these regions largely depend on rain-fed agriculture, and their livelihoods are vulnerable to environmental variability.

In the case of an emergency such as drought or flood, the most important mission for households and communities is to secure the food supply for survival. Figure 9.1 shows the social-ecological system of our study, which was adopted by the Research Institute for Humanity and Nature (RIHN) Social-Ecological Resilience

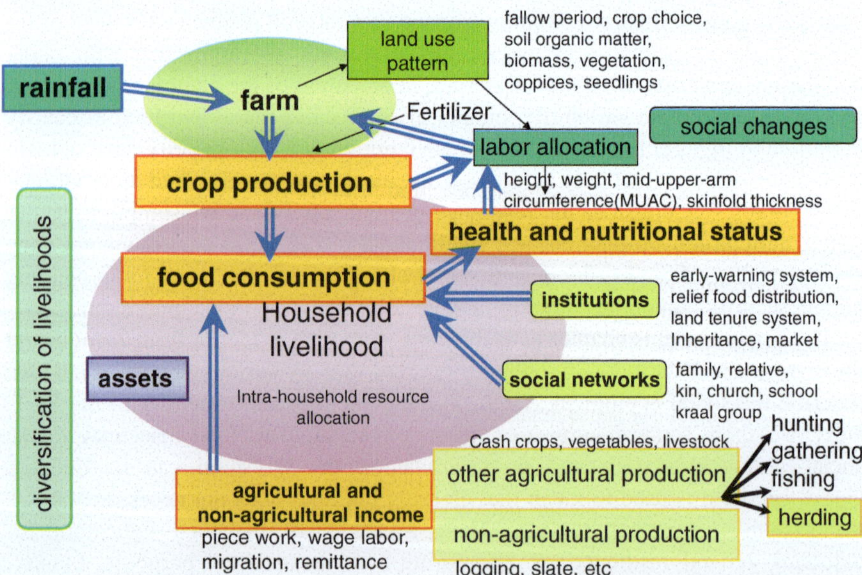

Fig. 9.1 Factors affecting resilience for food consumption levels among farm households in Zambia

Table 9.1 Concepts of resilience: from a narrow to a wide perspective

Resilience concept	Characteristic	Focus	
Engineering resilience	Return time, speed	Recovery	Short run
Ecological system resilience/social resilience	Buffer capacity	Robustness	Short run
Social-ecological resilience	Reorganization	Adaptive capacity, transformability, learning, reorganization	Short run and long run

From Folke (2006)

Project (Vulnerability and Resilience of Social–Ecological Systems) conducted from 2006 to 2011 (Umetsu 2011). This diagram indicates our research components, indicators, and factors affecting resilience and illustrates the linkage between rainfall, food supply, food consumption, health, and ecosystem services in drought-prone areas. Environmental variability such as rainfall and social changes (resilience *to* what) is shown in blue. Indicators are food supply, food consumption, food production, and health status (resilience *of* what) and are shown in orange. The connecting arrows show the working hypothesis of the project. Our purpose is to investigate the strength and weakness of the connection between these components, test the indicators of resilience, and verify factors and conditions for resilience.

Folke (2006) summarizes the concepts of resilience from a narrow to a wide perspective (Table 9.1). Engineering resilience is characterized by return time and speed. The focus of engineering resilience is recovery after disturbance of the system, whereas the main characteristics of ecological and social resilience are buffer capacity and robustness. The most recent and newly emerging concept of social-ecological resilience is characterized by reorganization and focuses on adaptive capacity, transformability, learning, and innovation. The engineering resilience of food security, the speed of recovery of food consumption, is reported in detail in Sakurai et al. (2011a, b), and the robustness of agro-ecological systems is reported in Shinjo et al. (2011). In this chapter we focus more on the recovery and adaptive capacity of social–ecological systems for food security.

Our three field sites were located in the Sinazongwe and Choma districts, Southern Province, Zambia (Fig. 9.2). The people in our sample villages are called Valley Tonga. After construction of the Kariba Dam in 1959, the Valley Tonga people suffered huge social and political shocks because of forced relocation from the valley bottom to the hill area (Colson 1960; Scudder 1962, 2010; Cliggett 2005). Site A (altitude 500 m) is close to Kariba Lake in an area of flat land; an old village was present before the dam was constructed, and the new village was relocated to site A after dam construction. Site B (altitude 700 m) is located in a mid-escarpment area with hilly farmlands. Residents of Site B were relocated to the current location during the 1990s. Site C is located at the highest altitude (1,050 m), at the edge of a plateau, and is an old village that existed before dam construction.

An intensive household survey was conducted to collect data; 16 sample farmers were selected from each site (A, B, and C). A total of 48 farm households were interviewed during the three cropping seasons, 2007/2008, 2008/2009, and 2009/2010.

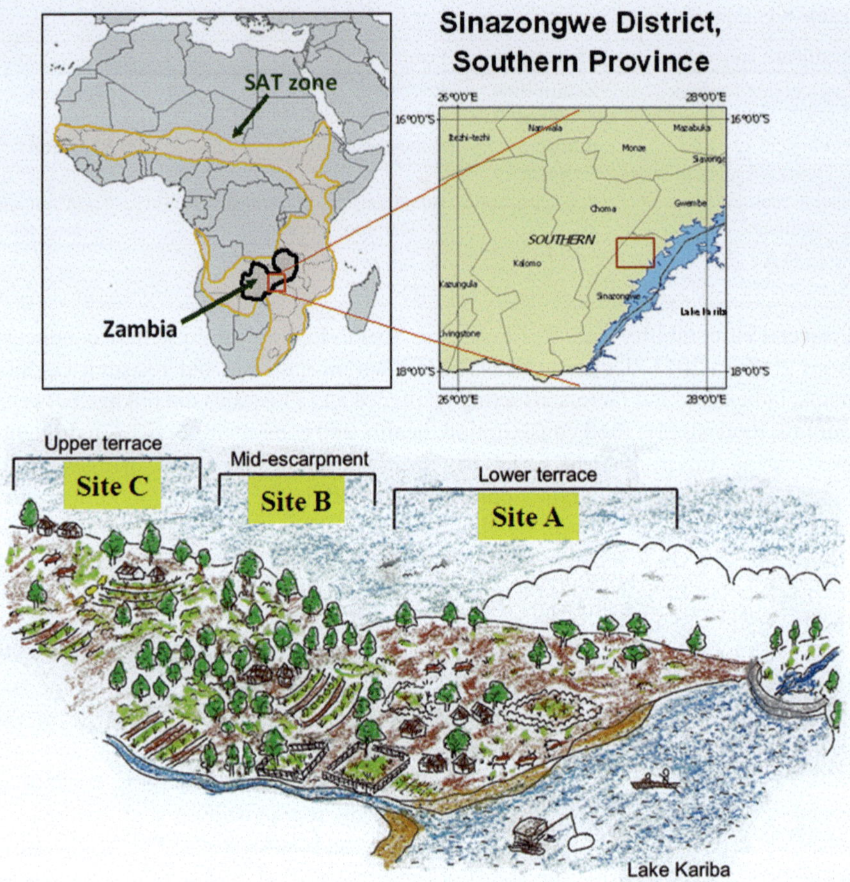

Fig. 9.2 Study sites in Sinazongwe District, Southern Province, Zambia (SAT: Semi-Arid Tropics)

Enumerators met sample farmers weekly and collected data on: (1) various social economic activities, including agricultural production, income, consumption, and time use; (2) body measurement, body weight, height, skin-fold thickness, and upper-middle arm circumference of household members; and (3) on-farm rainfall using rain gauges.

9.3 Threshold of Human System to Food Security against Climatic Variability

Threshold is defined as a certain level of key variables beyond which the system concerned moves to an alternative steady state and behaves in a different way with different feedbacks (Walker et al. 2006a; Gunderson 2003). The threshold provides important information about whether the system crossed the critical point and moved to an alternative state, i.e., regime shift. In systems ecology, this concept has been widely accepted and applied (Scheffer et al. 2001). However, when we consider the thresholds of human systems, the actual existence of such thresholds is less clear, because societies always have some feedback system to accept disturbance and avoid collapse, even in the face of catastrophic events such as natural disasters. Thus, we use the term quasi-threshold to mean that crossing the limits does not necessarily move the system to another regime. The following section discusses an environmental shock that occurred at our study sites and its consequences, i.e., quasi-thresholds in the context of food security under variable rainfall.

9.3.1 Rainfall Variability and Its Impacts on Agricultural Production

Environmental variability (e.g., rainfall variability) affects crop yield from a farmer's field, thus directly affecting food availability and consumption (i.e., survival of household). Historical rainfall data indicate that in Southern Province in Zambia the major droughts over the last 20 years occurred during the 1991/1992, 1994/1995, 2001/2002, and 2004/2005 cropping seasons. The production of maize, the major staple food in Zambia, as well as rural livelihoods has been directly affected by the level of precipitation. For example, the share of poverty in Southern Province increased from 79 % in 1991 to 86 % in 1993 immediately after the severe 1991/1992 drought (CSO 2007).

Figure 9.3 shows the daily mean and accumulated precipitation (mm) at Sites A, B, and C during the three cropping seasons, 2007/2008, 2008/2009, and 2009/2010. Rain usually begins in November and ends in April. We installed 48 rain gauges at our study sites to measure on-farm precipitation. Although our study sites are located in a drought-prone area of Zambia, precipitation during the three cropping seasons was much higher than the district's annual average (Kanno et al. 2011, 2013). On 29 December 2007, the Sinazongwe District, Southern Province, experienced heavy rainfall. The rain gauges we installed at Site A received 473 mm, on average, especially during the last week of December 2007, whereas the annual average rainfall in Sinazongwe District is 694.9 mm (Saeki et al. 2008). Although the two cropping seasons 2007/2008 and 2009/2010 were wet years, the rainfall patterns were quite different. During the 2007/2008 cropping season, heavy rain occurred in December, whereas during the 2009/2010 cropping season, heavy rain occurred in February. The heavy rain in December 2007 was associated with a La Nina year, and the heavy rain in February 2010 was associated with an El Nino year. Notably, farmers are facing not only seasonal variations but also annual variations in rainfall.

This heavy rainfall damaged the maize fields in the area. Among the three study sites, damage was the most severe at Site A. After the heavy rain, about 30 % of the damaged fields were abandoned, and only 54 % were replanted with maize.

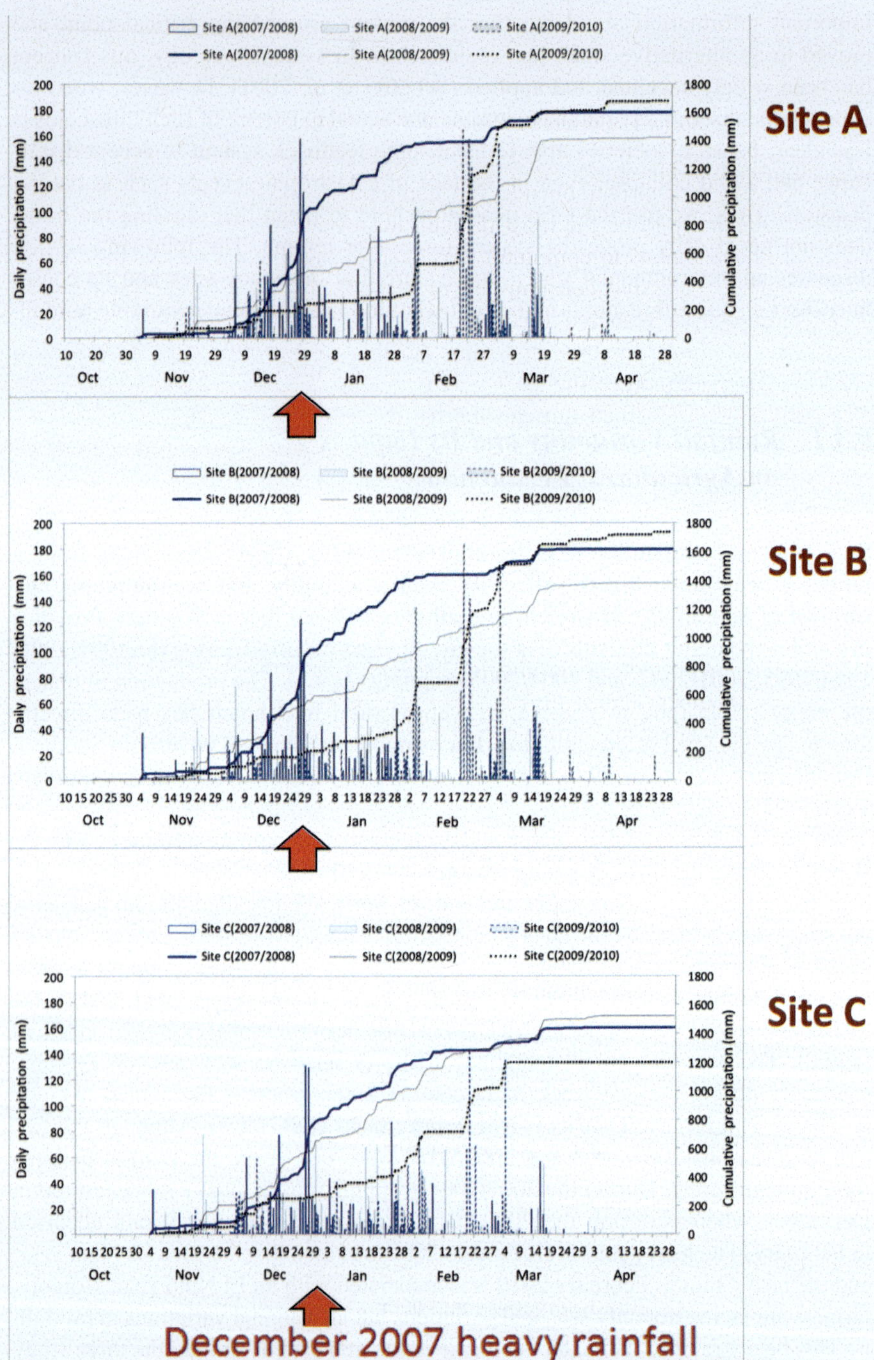

Fig. 9.3 Daily mean and accumulated precipitation (mm) at Sites A, B, and C during the 2007/2008 to 2009/2010 cropping seasons. From Kanno et al. (2011)

At Site C, switching from maize to sweet potato was a common practice. The topographical location of agricultural fields has been shown to mitigate climate variability (Yamashita et al. 2010; Miyazaki 2011a, b). This heavy rainfall resulted in a decline in maize production in 2008 and affected the food consumption, health, and nutritional status of household members.

9.3.2 Changes in Food Consumption and Quasi-Threshold

Food security means "the physical and economic access at all times to sufficient, safe and nutritious food to meet the dietary needs and food preferences for an active and healthy life" (IFAD 1996). Food security is closely related to the absolute concept of poverty, which emphasizes the lack of a given level of food. Thus, food insecurity is the situation where people fall below a pre-determined food security threshold (Bellu and Liberati 2005).

The Zambia Central Statistical Office (CSO) defines absolute poverty based on a minimum calorie intake, 2,094 kcal per adult per day (14,658 kcal/week), following the World Health Organization's recommendation. The World Bank uses a lower minimum food requirement standard, 1,334 cal per adult per day (12,418 kcal/week) (World Bank 2007). In monetary terms estimated from the minimum food basket, the extremely poor group is defined by an income level below 65,710 Zambian Kwacha (ZMK) per adult per month in 2006. This income level considers only food expenditure and, thus, indicates the level of food insecurity. The moderately poor group is defined by an income level above 65,710 ZMK but less than 93,872 ZMK per adult equivalent per month in 2006. This income level includes food and non-food expenditures. An income level above 93,872 ZMK per adult per month is considered non-poor.

Figure 9.4 shows the changes in average food consumption level per week per adult equivalent at Site A from November 2007 to December 2009. The red lines indicate the quasi-threshold of food security, in this case a minimum calorie intake of 14,658 kcal per adult per week and 12,418 kcal per adult per week. This is the quasi-threshold because it is a pre-determined food security threshold and crossing it does not necessarily cause catastrophic events or regime shifts. Our survey revealed the dynamic changes in household food consumption levels with various environmental and socio-economic changes. After the heavy rain in December 2007, calorie intake decreased from 12,000 to 4,000 kcal/week. The level of food consumption did not recover even after the following harvest season. Because of the heavy rain, about 34 % of the maize fields at Site A were damaged (20 % at all three sites) and maize production was reduced (Miyazaki 2011a, b). Reduced production caused an extended lean period, longer than in normal years. In this region, the maize stock does not usually last until the next harvest season, and maize stock of relatively poor households starts to deplete as early as October, which is the beginning of the next planting season. A similar nutritional study in Burkina Faso also confirmed that the calorie intake of household members was

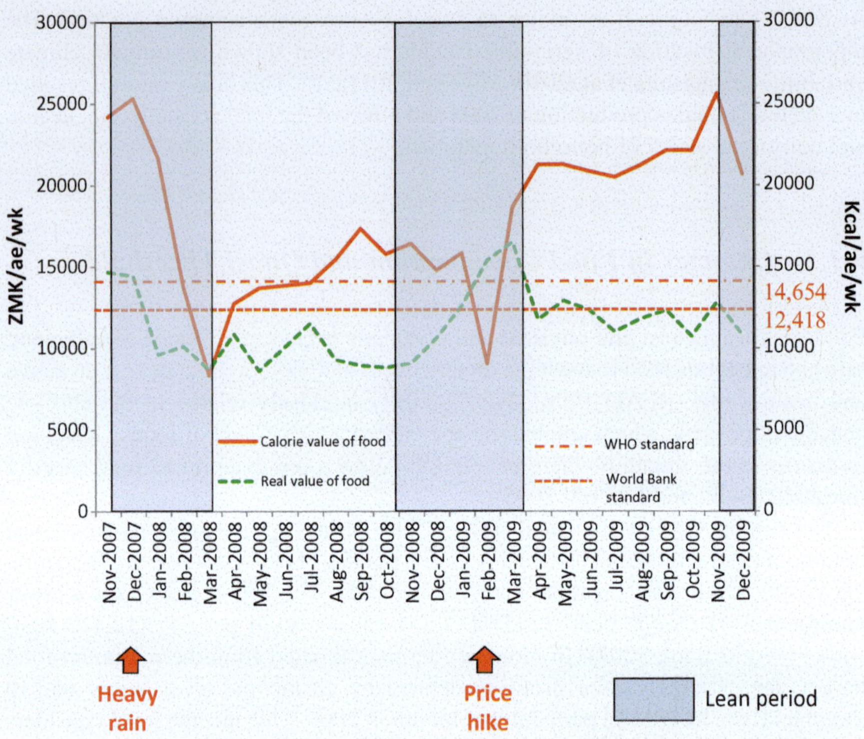

Fig. 9.4 Average food consumption per adult equivalent (ae) per week at Site A. From Sakurai et al. 2011a, b

significantly reduced from 3,000 kcal/day to about 1,500 kcal/day during the food shortage months (Ishimoto 2010).

During the 2008/2009 cropping season, the maize price increased not only locally but also nationally. The retail maize price in Choma, the nearest industrial city where maize mielie meal is produced, soared from 1,140 ZMK in January 2008 to 1,920 ZMK in January 2009, a 68 % price increase. This price hike and local maize shortage due to heavy rain during the previous cropping season in 2007/2008 also affected the local maize market significantly and increased the price further. The price hike had a particular implication for poor farmers, whose harvests were already depleted and who had to secure cash to purchase food.

9.3.3 Changes in BMI Over Time and Quasi-Threshold

Household welfare is affected not only by the quantity of food household members consume but also by the food quality, food diversity, and how the food is prepared. The quantity of food consumed as an input variable into the production of

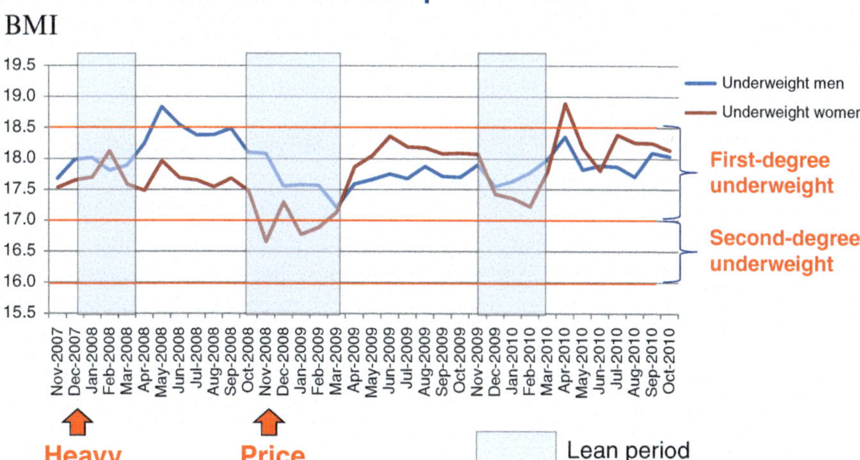

Fig. 9.5 Body mass index (BMI) and nutritional quasi-threshold of BMI-based first- and second-degree underweight men and women. From BMI index provided by H. Kubo

household welfare is a reasonable indicator of household welfare in the absence of other more detailed data. Nutritional status has a closer link to all four dimensions of food security, i.e., availability, accessibility, utilization, and stability, and can be considered as an important indicator of food security. While nutritional status is a better surrogate of food security status than food consumption level because of its ability to capture a more complete aspect of food security, the nutritional status variable has a disadvantage compared to the food quantity variable in that it is less sensitive to immediate changes in household welfare.

For this purpose, a body mass index [BMI = body weight (kg)/height (m^2)] of 18.5 was used as another quasi-threshold, in addition to calorie intake. BMIs were calculated using height and weekly measured body weight data for each household member (Yamauchi et al. 2011). A BMI below 18.5 is categorized as first-degree underweight, and a BMI below 16.0 is categorized as second-degree underweight. Yamauchi et al. (2011) found that most adult household members have a good nutritional status, with BMIs in the normal range (between 18.5 and 25). A small fraction of the sampled population was classified as underweight by the BMI standard.

Figure 9.5 shows BMIs and the nutritional quasi-threshold for the underweight group. This figure, in addition to the food consumption level in Fig. 9.4, shows dynamic changes in the nutritional status of village households in terms of BMI and highlights the impact of climate shock on the health and nutritional status of the vulnerable. Figure 9.4 clearly indicates that the nutritional status of household members is seldom above the quasi-threshold level of BMI 18.5 and fluctuates throughout the year depending on various external factors and internal mechanisms which adjust and stabilize body weight. At Site A, after the heavy rain in December 2007, some farmers replanted maize, which increased the time allocation for agricultural labor and resulted in declines in body weight. Further, due to the price hike mentioned above, food consumption and calorie intake decreased during the lean period and BMI declined sharply.

Changes in the livelihoods of farmers and the underlying reasons, both environmental and socio-economic, must be recognized to understand the overall context. From a resilience perspective, it is important to consider how household members recover from shock. After the heavy rain in December 2007, 20 % of maize fields were damaged at all three sites. Among all study sites, Site A received most severe damage (34 % of maize fields were damaged). Some farmers replanted maize, some switched to sweet potato cultivation, and some engaged in selling livestock to secure cash income. Our analysis shows that when total income decreases, staple food consumption (mainly maize) decreases, but vegetable consumption decreases to a lesser extent (Kitsuki and Sakurai 2011). This is because cash is needed for households to purchase maize but is not needed to collect wild fruits and vegetables, which indicates the importance of cash income during food shortage periods for staple food consumption.

Farmers living near the poverty line often temporarily cross this food poverty and nutritional quasi-threshold. For the transitional food poverty, we must understand the causes and mechanisms of this dynamic change and develop countermeasures which enhance household capacity to adapt and cope to mitigate food insecurity and its impact for faster recovery.

9.4 Assessing General Resilience at a Higher Scale

Social–ecological systems are linked across spatial and temporal scales. Resilience of an existing scale depends on what happens at a scale above and below it (Walker et al. 2006a). Households, local communities, and regional and global markets intricately influence, and are influenced by, each other. Although it is beyond the scope of this chapter to test this cross-scale dynamic relationship empirically, decisions made at one level will influence the others and vice versa. Downward resilience at a macro level and upward resilience at a micro level can enhance or erode resilience at the scale under study. For each study site, household is the lower scale; the district and higher level is the site's higher scale. Upward resilience at the household level was examined with respect to the heavy rainfall shock. Our attention will now shift to downward resilience at a macro level.

The resilience assessment in previous sections involved identifying threshold effects after flood and heavy rain perturbations. This type of specified resilience assessment is necessary because the threat has known threshold effects. Farmers can fall into a poverty trap if the climate shock is pervasive and prolonged. Focusing on climate risks alone is not sufficient, because farm households face other risks besides climate. Excessive attention to develop specified resilience may result in an unintended loss of general resilience to other new shocks. Here, we approach general resilience to food insecurity by assessing the general capacity of households to absorb shocks, self-reorganize, learn and innovate in combination of assessing four dimensions of food security.

Resilience to food insecurity is a multi-faceted concept and is not readily observable. Direct assessment of resilience for monitoring and management in a non-quasi-experimental setting is challenging. Researchers at the Food and Agriculture Organization of the United Nations (FAO) innovatively proposed an indirect method to assess resilience by modeling resilience to food insecurity as a composite of latent variables (Alinovi et al. 2010a, b). A multi-stage latent variable model is used to estimate resilience scores based on a two-stage factor analysis estimation. The proposed resilience assessment method provides not only a quantification of resilience for resilience monitoring but also an identification of causes of being less resilient for resilience management. The resilience estimation method involves resilience component identification, resilience component function specification and estimation, and resilience score estimation as a weighted sum of its components.

Here, we assert that resilience to food insecurity (R) is a function of the following factors (Alinovi et al. 2010a):

$$R = f(Income, Asset, SafetyNet, BasicServices, AgroTechno, AdaptiveCapacity, Stability) \tag{9.1}$$

where *Income* is having income and access to food at all times; *Asset* is having agricultural and non-agricultural assets such as house, land, and livestock for productive use and for a cushion against shocks; *SafetyNet* is having access to social safety nets such as food aid, food for work, and remittances; *BasicServices* is having access to basic services such as clean drinking water, sanitation, health care, electricity, etc.; *AgroTechno* is having sound agricultural and technological practices such as access to extension and veterinarian services and to agricultural inputs; *AdaptiveCapacity* is having the capacity to adapt to shocks such as education and diversity of income sources; and *Stability* is having stability in all of the above components over time.

The components of resilience to food insecurity are specified based on the four pillars of food security, i.e., food production or availability, food access, food utilization, and stability. Although agricultural and technological practices determine food production, income influences access to food. Access to basic services such as drinking water and health care services can be loosely linked to food utilization. An absence of shocks was an indicator of stability in food supply, food access, and utilization. When shocks were unavoidable, assets, safety nets, and adaptive capacity allowed individuals and households to mitigate the impact of the shocks and smooth food consumption levels.

9.4.1 *Defining Indicator Variables*

The determinants of the resilience components are specified as follows:

Component	Indicator
Income and food access	Per capita income (continuous)
	Per capita consumption (continuous)
	Household perception of food security (scale of 1 to 3)
Asset	Non-agricultural asset values per capita (continuous)
	Agricultural asset values per capita (continuous)
Social safety net	Amount of remittance and transfer received (continuous)
Access to basic services	Access to clean water (dummy)
	Access to credit (dummy)
	Access to electricity (dummy)
	Access to telephone (dummy)
	Distance to public transportation (continuous)
	Distance to basic school (continuous)
	Distance to health center (continuous)
	Perception of school quality (dummy)
Agricultural practices and technology	Use of chemical fertilizers (dummy)
	Use of manure (dummy)
	Use of pesticides (dummy)
	Use of veterinarian services (dummy)
	Use of irrigation (dummy)
Adaptive capacity	Number of household income-generating activities (count)
	Adult household members in employment activities (%)
	Average years of education of adult household members (continuous)
	Non-food expenditure (%)
	Basic coping capacity defined as an inverse of the severity-adjusted coping strategy index (coping capacity is set to 1 if the household has not used any coping strategies) (%)
Stability (multiplying each indicator by −1 to change an instability indicator to a stability indicator)	Self-assessed income stability (categorical)
	Household members that lost job in the last 12 months (dummy)
	Household members that fell ill during the last 2 weeks (%)
	Share of transfer (%)

Data used in this analysis are based on Zambia's 2004 Living Condition Monitoring Survey (LSMS), 2004. The survey is nationally representative and was conducted by the Central Statistical Office (CSO). Sample size was 19,340 households.

9.5 Results of Resilience Indicator Analysis

The resulting resilience scores and scores of its components were normalized with zero mean and unit variance. A profile of resilience to food insecurity was as expected. Female-headed households were more vulnerable than male-headed households (−0.1298947 and 0.0390073, respectively). Resilience to food insecurity varied by household-head age groups in an inverted U shape (see Fig. 9.6) and peaked at a relatively young age (26–35 years). This was not surprising considering that Zambia's life expectancy at birth in 2009 was 46.3 (World Bank 2011) and age at retirement was 55. Those aged beyond 55 form the most vulnerable group.

The agricultural sector was generally the least resilient group, followed by the construction and manufacturing sectors; these groups are largely comprised of low-skilled labor or the urban poor (Fig. 9.7). Households working in the utilities and financial sectors appeared to be the most food secured. Within the agricultural

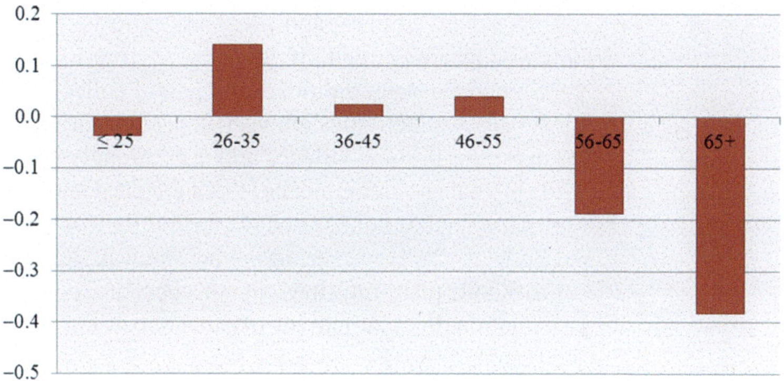

Fig. 9.6 Resilience scores by age group of household head

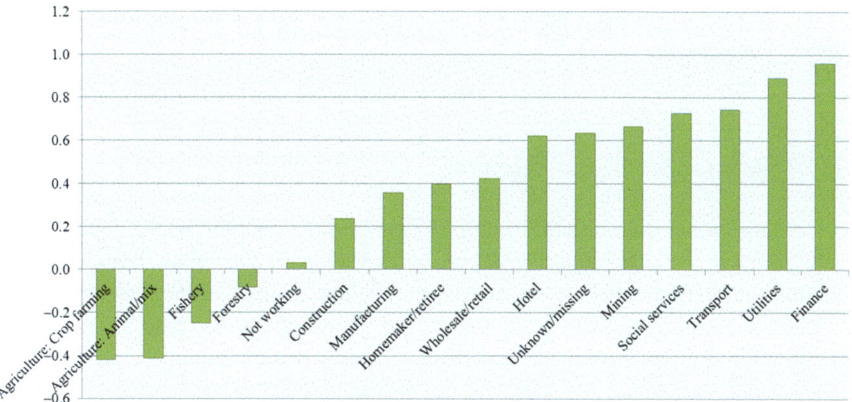

Fig. 9.7 Resilience scores by occupational industry of household head

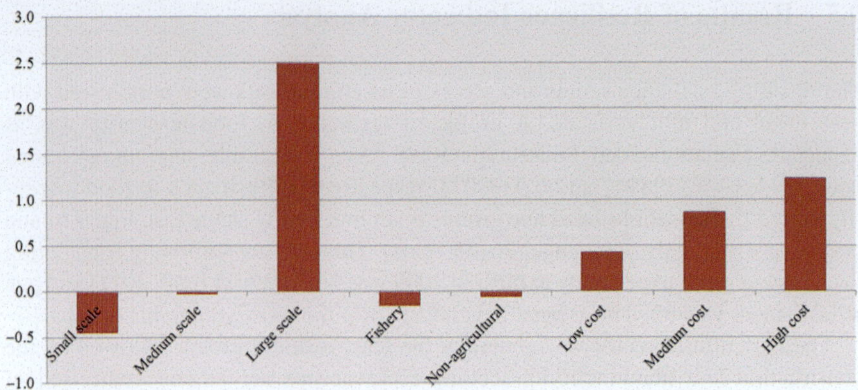

Fig. 9.8 Resilience scores by household stratum

Fig. 9.9 Map of resilience scores by province

sector, large-scale commercial farmers were the most resilient overall, and small-scale farmers were the least resilient overall (Fig. 9.8). Urban households in locations ranging from low- to high-cost residential areas were generally more resilient to food insecurity than their rural counterparts. A map of resilience indicators within the provincial and district boundaries of Zambia is presented in Fig. 9.9 and shows that the relatively more urbanized provinces, such as Lusaka, Copperbelt, and Southern, are more resilient to food insecurity than the less developed provinces. Table 9.2 details the suggested causes of resilience. The

Table 9.2 Resilience scores and components by province

Component	Province								
	Lusaka	Copperbelt	Southern	Central	Northwestern	Luapula	Northern	Eastern	Western
Resilience	0.647	0.414	−0.011	−0.152	−0.219	−0.264	−0.298	−0.303	−0.385
Access to basic services	0.720	0.572	0.001	−0.309	−0.357	−0.171	−0.480	−0.221	−0.480
Adaptive capacity	0.701	0.391	−0.013	−0.094	−0.197	−0.326	−0.269	−0.431	−0.303
Access to food	0.203	0.097	−0.103	−0.018	0.102	−0.099	0.043	−0.092	−0.287
Non-agricultural asset	0.082	0.010	−0.054	−0.012	−0.090	−0.116	−0.101	−0.085	−0.109
Social safety net	0.044	0.009	0.093	−0.098	−0.139	−0.143	−0.091	−0.039	−0.121
Agricultural asset	−0.005	−0.008	−0.007	0.036	−0.008	−0.008	−0.008	−0.008	−0.008
Agro-technological practice	−0.286	−0.264	0.212	0.173	−0.243	−0.261	−0.143	0.822	−0.173

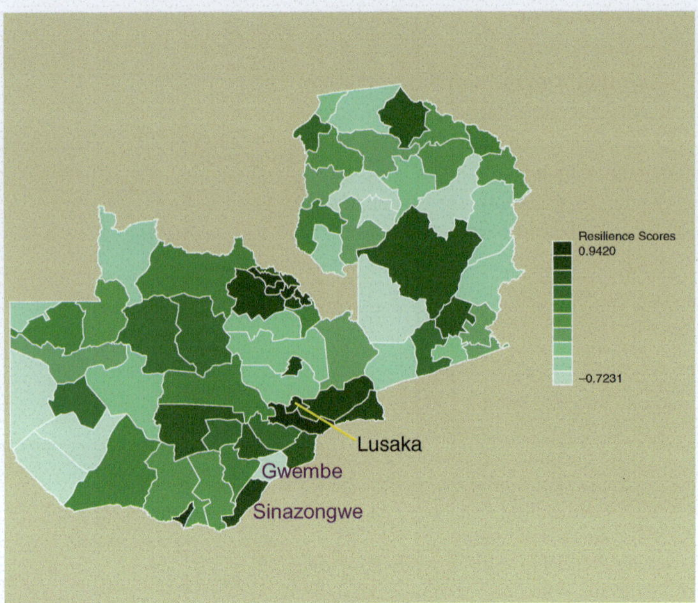

Fig. 9.10 Resilience map at the district level, Zambia, 2004

more resilient provinces tended to have better access to basic services, greater adaptive capacity, greater access to food and income, more non-agricultural assets, and better access to social safety nets. Figure 9.10 shows that the Sinazongwe District in Southern Province, where our study sites were located, is relatively more resilient than all districts except Livingstone, the former capital city, because of better access to basic services and greater adaptive capacity and income-generating potential (table not shown but available upon request). Gwembe, a district with a relatively poor infrastructure, is the least resilient in Southern Province (see Fig. 9.10). Resilience in the provinces and districts is consistent with our findings at the community level, i.e., the majority of the population is in good health and nutritional status despite the rainfall shock.

The resilience framework can also be used to shed light on poverty and consumption vulnerability classification. Here, vulnerability to food insecurity is defined as the probability of the future consumption of households falling below a minimum consumption threshold and is estimated econometrically using Chaudhuri's method (Chaudhuri et al. 2002). A household is categorized as being a member of one of the four possible groups, i.e., the vulnerable poor, the vulnerable non-poor, the non-vulnerable poor, and the non-vulnerable non-poor. In Zambia, 70 % of households were considered vulnerable, and of those, 43 % were currently poor and 27 % were currently not poor. Of the 30 % non-vulnerable households, 11 % were currently poor and the remaining 19 % were non-poor.

Table 9.3 Poverty, vulnerability, and resilience to food insecurity

	Mean score			
	Non-poor		Poor	
Component	Non-vulnerable	Vulnerable	Non-vulnerable	Vulnerable
Vulnerability	0.180	0.847	0.188	0.894
Resilience	0.952	−0.051	0.219	−0.438
Adaptive capacity	1.014	−0.187	0.505	−0.451
Access to basic services	0.925	0.032	0.054	−0.438
Access to food	0.408	0.091	−0.097	−0.211
Social safety net	0.329	0.019	−0.177	−0.196
Non-agricultural asset	0.164	−0.016	−0.118	−0.128
Agricultural asset	−0.006	0.008	−0.008	−0.008
Stability	−0.187	−0.070	−0.026	0.132
Agro-technological practice	−0.222	0.225	−0.221	0.013

What makes the non-poor vulnerable was their poorer access to basic services, lower adaptive capacity, lower income-generating capacity, and lower access to social safety nets compared to the non-vulnerable non-poor. Their higher agricultural asset holdings, higher employment stability, and greater use of agricultural inputs and agricultural technology suggest that the vulnerable non-poor are likely to be rural medium-scale farmers.

Compared to the vulnerable poor, the non-vulnerable poor had greater adaptive capacity, easier access to basic services, and better income-generating potential. Lower employment stability and lower use of agricultural inputs and technology probably indicates that the non-vulnerable poor were the urban poor (Table 9.3).

9.6 Conclusion and Recommendations

In the SAT, people's livelihoods are vulnerable to environmental variability. The SAT includes Sub-Saharan Africa and South Asia, where the absolute number and proportion of people who are extremely impoverished will remain large for some time to come. People in these regions largely depend on vulnerable rain-fed agriculture, and food security, livelihood resilience, and poverty reduction are critical issues. As ex-ante and ex-post risk-coping strategies, the capacity of diversified access to resources is one of the important conditions for resilience. Access to resources is facilitated through a transfer and/or substitution of livelihood from agriculture to livestock, agriculture to non-agriculture, market, social organization, and institution, as well as social network. Rural households and communities in Africa are facing risks not only from natural disasters but also from social and economic changes, such as international price hikes of cash crops, political transition, and changes in land tenure systems and agricultural policies.

Empirical evidence shows that the food security of farmers is threatened by temporally and yearly environmental shock. It is commonly found that some poverty groups often cross the quasi-threshold of food security.

The analysis of large samples revealed that small-scale farmers are where the problem of low resilience to food insecurity exists. They are also a host to solutions for poverty reduction and community food security improvement. Building resilience to food insecurity is a complex policy issue because it involves a large and intertwined set of public policies, including education, health, population, migration, infrastructure, institution, agriculture, social insurance, and safety nets. General resilience to food insecurity can be enhanced by building adaptive capacity, increasing access to basic services, providing a stable economic environment and income-generating capacity, and increasing assets. Adaptive capacity can be enhanced by increasing human capital through education, reducing adult mortality to maintain a high percentage of working adults in the population, allowing flexible immigration policies for effective labor supply, and diversifying the economy to provide diversified livelihood opportunities. Government should also focus their resources on providing social and physical infrastructure such as social insurance, social safety nets, health care facilities, electricity, roads, and public transportation to help markets operate more efficiently. The public should also invest in agricultural research and development to improve seed crops by focusing on shortening maturity times even further to reduce exposure to climatic risks. The sustainable provision of extension services through collaborations with the private sector, non-governmental organizations, and farmers is essential to increase productivity in the agricultural sector.

Acknowledgements This is a contribution from the Social-Ecological Resilience Project (E-04, Vulnerability and Resilience of Social–Ecological Systems), administered by the RIHN, and JSPS Grant-in-Aid for Scientific Research (B) Program No. 23310027. We acknowledge Mr. Harutaka Kubo, Graduate School of Health Sciences, Hokkaido University, for his excellent assistance with the BMI calculations.

References

Alinovi L, D'Errico M, Mane E, Romano D (2010) Livelihoods strategies and household resilience to food insecurity: an empirical analysis to Kenya. Paper prepared for Societa' Italiana Di Economia Agraria XLVII Convegno di Studi "L'agricoltura oltre le crisi", Campobasso

Alinovi L, Mane E, Romano D (2010b) Measuring household resilience to food insecurity: application to Palestinian households. In: Benedetti R, Piersimoni F, Bee M, Espa G (eds) Agricultural survey methods. Wiley, West Sussex

Bellu LG, Liberati P (2005) Impacts of policies on poverty: the definition of poverty. Food and Agriculture Organization of the United Nations (FAO), Rome

Berkes F, Folke C (eds) (1998) Linking social and ecological systems: management practices and social mechanisms for building resilience. Cambridge University Press, Cambridge

Berkes F, Colding J, Folke C (eds) (2003) Navigating social-ecological systems. Cambridge University Press, Cambridge

Chaudhuri S, Jalan J, Suryahadi A (2002) Assessing household vulnerability to poverty from cross-sectional data: a methodology and estimates from Indonesia. Department of Economics, Columbia University, New York

Cliggett L (2005) Grains from grass: aging, gender, and famine in rural Africa. Cornell University Press, Ithaca

Colson E (1960) The social organization of the Gwembe Tonga. Manchester University Press, Manchester

CSO (2004) Living condition monitoring survey. Survey in December 2004. Central Statistical Office, Lusaka

CSO (2007) Living conditions monitoring survey report 2006. Living Condition Monitoring Branch, Central Statistical Office, Lusaka, Zambia

Folke C (2006) Resilience: the emergence of a perspective for social-ecological systems analyses. Global Environ Change 16:253–267

Gunderson LH (2003) Adaptive dancing: interactions between social resilience and ecological crises. In: Berkes F, Colding J, Folke C (eds) Navigating social-ecological systems: building resilience for complexity and change. Cambridge University Press, Cambridge

Gunderson LH, Allen CR, Holling CS (eds) (2006) Foundations of ecological resilience. Island Press

Holling CS (1973) Resilience and stability of ecological systems. Annu Rev Ecol Syst 4:1–23

ICRISAT (2010) Strategic plan to 2020: inclusive market-oriented development for smallholder farmers in the tropical drylands. International Crops Research Institute for the Semi-Arid Tropics, Hyderabad

ICSU (2010) Earth system science for global sustainability: the grand challenges. International Council for Science, Paris

IFAD (1996). Household food security: Implications for policy and action for rural poverty alleviation and nutrition, IFAD Paper for the World Food Summit. International Fund for Agricultural Development, Rome

Ishimoto Y (2010) Food consumption system as a survival strategy in the semiarid area: an empirical analysis on agropastoralists in the Sahel. J Arid Land Stud 20(2):85–95

Kanno H, Shimono H, Sakurai T, Yamauchi T (2011) Analysis of meteorological measurements made over the 2008/2009 rainy season in Sinazongwe District, Zambia. Vulnerability and Resilience of Social-Ecological Systems, FR4 Project Report, pp 123–135

Kanno H, Sakurai T, Shinjo H, Miyazaki H, Ishimoto Y, Saeki T, Umetsu C, Sokotela S, Chiboola M (2013) Indigenous climate information and modern meteorological records in Sinazongwe District, Southern Province, Zambia. J Agric Res Quart 47(2):101–201

Kitsuki A, Sakurai T (2011) Seasonal consumption smoothing in rural Zambia. J Rural Econ Spec Issue 2011:380–384

Levin SA (1999) Fragile dominion: complexity and the commons. Perseus Books, Reading

Levin SA, Barrett S, Aniyar S, Baumol W, Bliss C, Bolin B, Dasgupta P, Ehrlich P, Folke C, Gren I-M, Holling CS, Jansson A-M, Jansson B-O, Martin D, Mäler K-G, Perrings C, Sheshinsky E (1998) Resilience in natural and socioeconomics systems. Environ Dev Econ 3(2):222–234

Mäler K-G (2008) Sustainable development and resilience in ecosystems. Environ Res Econ 39:17–24

Miyazaki H (2011a) Adaptation and coping behavior of farmers during pre- and post-shock periods. Vulnerability and Resilience of Social-Ecological Systems, FR4 Project Report, pp 164–170

Miyazaki H (2011b) Adaptive and coping behaviors with rainfall fluctuation by small-scale farmers in Southern Province of Zambia. Poster presented at Resilience 2011, Arizona State University, Tempe

Perrings C (2006) Resilience and sustainable development. Environ Dev Econ 11:417–427

Ryan JG, Spencer DC (2001) Future challenges and opportunities for agricultural R&D in the semi-arid tropics. Patencheru 502 324, Andhra Pradesh, India, International Crops Research Institute for the Semi-Arid Tropics

Saeki T, Kanno H, Miyazaki H, Shinjo H (2008) Meteorological observation in Southern Province, Zambia. In: Proceedings of the Meteorological Society of Japan 2008 fall meeting, Sendai, 19–21 November, p 473 (in Japanese)

Sakurai T, Nasuda A, Kitsuki A, Miura K, Yamauchi T, Kanno H (2011a) Measuring resilience of household consumption – the case of the Southern Province of Zambia. J Rural Econ 393–400 (in Japanese)

Sakurai T, Nasuda A, Kitsuki A, Miura K, Yamauchi T, Kanno H (2011b) Vulnerability and resilience of households: the case of Zambia. Econ Rev 62(2):166–187 (in Japanese)

Scheffer M, Carpenter S, Foley JA, Folke C, Walker B (2001) Catastrophic shifts in ecosystems. Nature 413:591–596

Scudder T (1962) The ecology of the Gwembe Tonga. Manchester University Press, Manchester

Scudder T (2010) Global threats, global futures: living with declining living standards. Edward Elgar Publishing, UK

Shinjo H, Ando K, Kuramitsu H, Miura R (2011) Effects of clearing and burning on soil nutrients and maize growth in a Miombo woodland in eastern Zambia. Vulnerability and Resilience of Social-Ecological Systems, FR4 Project Report, pp 15–32

Troll C (1965) Seasonal climates of the earth. In: Rodenwaldt E, Jusatz HJ (eds) World maps of climatology, 2nd edn. Springer, Berlin

Umetsu C (2011) Resilience of rural households in semi-arid tropics: a linkage of social-ecological systems. Vulnerability and resilience of social-ecological systems, FR4 Project Report, pp 8–14

UNDP, UNEP, WB, WRI (2008) World Resources 2008: roots of resilience-growing the wealth of the poor. World Resources Institute, Washington, DC

Walker B, Holling CS, Carpenter SR, Kinzig A (2004) Resilience, adaptability and transformability in social-ecological systems. Ecol Soc 9(2):5

Walker B, Gunderson L, Kinzig A, Folke C, Carpenter S, Schultz L (2006a) A handful of heuristics and some propositions for understanding resilience in social-ecological systems. Ecol Soc 11(1):13

Walker B, Salt D, Reid W (2006b) Resilience thinking: sustaining ecosystems and people in a changing world. Island Press, Washington, DC

World Bank (2007) Zambia poverty and vulnerability assessment. Report No. 32573-ZM. World Bank, Washington, DC

World Bank (2011) World development indicators. World Bank, Washington, DC

Yamashita M, Miyazaki H, Ishimoto Y, Yoshimura M (2010) Multi-temporal and spatial data integration for understanding the livelihood in village level. Int Arch Photogram Rem Sens Spatial Inform Sci XXXVIII(8):827–830

Yamauchi T, Kubo H, Kon S, Thamana L, Sakurai T, Kanno H (2011) Growth and nutritional status of Tonga children in rural Zambia–longitudinal growth monitoring over 26 months. Vulnerability and Resilience of Social-Ecological Systems, FR4 Project Report, pp 104–110

Chapter 10
Changes in Resource Use and Subsistence Activities Under the Plantation Expansion in Sarawak, Malaysia

Yumi Kato

Abstract This chapter considers the effect of plantation development on human resource use and subsistence activities in Sarawak, Malaysia. With the demand for palm oil, natural forest has rapidly changed into plantations in tropical areas in recent decades. While the large-scale plantations expand dramatically, it is unclear how their resource use and subsistence activities change after plantations spread. This chapter examines the alteration of subsistence activities, including, shifting cultivation, hunting, gathering wild plants, and cash crop cultivation.

I found that shifting cultivation has not much changed in terms of work processes, cultivation areas, and crops. However, an increasing number of families have recently cultivated their paddy fields along roads. As for hunting, before plantation expansion the local population hunted various animals in the natural forest, but after it they mostly chase wild boar in the plantations at night. They used diverse wild plants of the natural forest before plantations spread, but these wild plants have sharply decreased around plantations. After the expansion of road networks and oil palm plantations, people now have new opportunities to cultivate oil palms as smallholders. This chapter illustrates how people try to continue the conventional way of livelihood even under the plantation expansion. They also adapt flexibly to the new environment by using resources available around the plantations.

Keywords Land use change • Livelihood • Oil palm plantation • Resource use • Riverine society

Y. Kato (✉)
Hakubi Center for Advanced Research, Kyoto University, Kyoto, Japan
e-mail: yumi.katou@gmail.com

S. Sakai and C. Umetsu (eds.), *Social–Ecological Systems in Transition*, Global Environmental Studies, DOI 10.1007/978-4-431-54910-9_10, © Springer Japan 2014

10.1 Introduction

The biodiversity of Borneo is one of the richest in the word. It has numerous endemic plants and animal species, including many that are threatened. This rich biodiversity has supported the subsistence activities of the local people for a long time. It has also been the base of the country's economic development, which depended on the trade in various forest products, such as sago palm, rattan, iron wood, bird's nests, and rubber, until the early twentieth century (Ooi 1995). At that time, the local people were active as collectors of or traders in such products along the river basins. However, export of logged timber became one of the main economic industry of the country from the 1960s. Since the 1980s, oil palm and acacia (*Acacia mangium*) plantations have expanded. At the same time, the social environment has also changed because of the demand of works in urban areas and expansion of road networks into inland Borneo. The local people have moved from riverside longhouses to urban areas and new settlements along roads.

Oil palm plantations have rapidly expanded, but few studies discuss the changes in the resource use and subsistence activities of the local population after plantations spread. This study seeks to understand these effects. First, I outline the land use changes and expansion of oil palm plantations in Sarawak, Malaysia. Second, I describe the change of people's living environments caused by expansion of road networks. Third, I investigate the alterations in subsistence activities, focusing on shifting cultivation, hunting, gathering wild plants, and oil palm smallholdings.

10.2 Land Use Changes on Inland Sarawak

In this section I describe the historical changes in land and resource use in Sarawak. Resource use in Sarawak has changed in the long term. Before the early twentieth century, the country's economic activity depended on the trade of various non-timber forest products, which had begun in the tenth century. Various forest products, such as rhinoceros horns, bezoar stones, bird nests, camphor, and incense wood, were exported to China in the Sung period of the tenth century (Tagliacozzo 2005). Until 1910 the trade in non-timber forest products was the main economic resource of Sarawak (Ooi 1995). The sago palm was the main export throughout the nineteenth century, followed by bezoar stones, gutta percha (*Palaquium gutta*), Indian rubber (*Ficus elastica*), and antimony. After 1900, wild rubber became the top export product, while others, such as gutta percha, jelutong (*Dyera costulata*), rattan, dammar, illipe nuts, and honey, were also exported. However, between 1920 and 1940, petroleum became the dominant export. The main export commodities in the first half of the 1940s were petroleum, rubber, pepper, timber, sago palm, and jelutong. The trade in non-timber forest products was still important until the middle of the twentieth century, although the ratio of timber grew in its final decades.

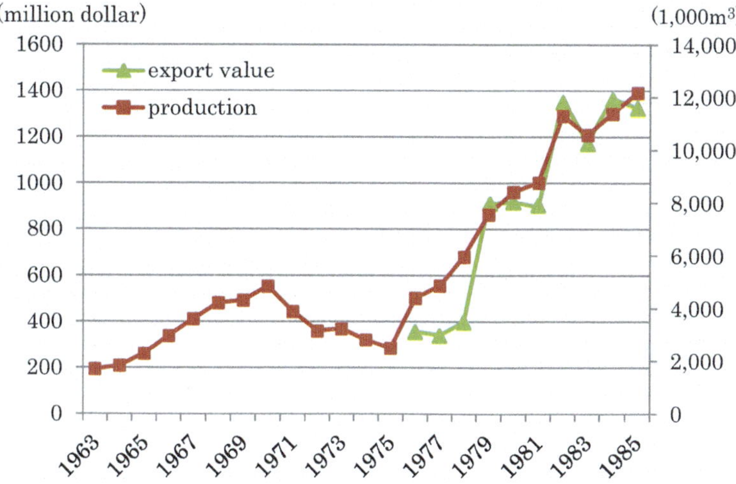

Fig. 10.1 Timber production and export in Sarawak. *Source*: Department of Statistics Sarawak, Forest Department Sarawak

Timber, such as Borneo iron wood (*Eusideroxylon zwageri*), was already being exported in the nineteenth century, but it accounted for only 2–3 % of the export revenue of Sarawak in the first half of the twentieth century (Ooi 1995; Smythies 1963). However, commercial logging has expanded toward the inland hilly areas since the 1960s with the introduction of heavy machinery. This initiated remarkable economic development, especially in the 1980s (Fig. 10.1). Timber became the second largest export commodity in 1980s, following petroleum. In 1983, Malaysian timber accounted for 58 % of the total exported timber of the world market. The export of logged timber was the most important economic revenue of the latter half of the twentieth century.

During the 1990s, oil palm production became the main agricultural product of Sarawak. The spread of oil palm cultivation has been astonishing in insular Southeast Asia over the past few decades. Previously found in Peninsular Malaysia and in Sumatra, Indonesia, the crop currently has one of its frontiers in Borneo (De Koninck 2011). Cultivation flourished from the 1970s to the 1980s in Peninsular Malaysia, where, with population increase and industrial development, land for oil palm cultivation has been insufficient. In Malaysia overall, the center of oil palm production moved from the peninsula to the state of Sabah in the 1990s and then to Sarawak by the latter half of that decade. The first oil palm plantation in Sarawak was established in the Miri Division in 1969. The publishing of the *Konsep Baru* (New Concept) also promoted the active use of native land for oil palm plantations through joint ventures of the government and private enterprise (Majid Cooke 2002; Ngidang 2002; Cramb and Sujang 2011). Figure 10.2 shows the expansion of estimated land for oil palm in Malaysia. Plantation has rapidly spread from the 1980s. In 2005, oil palm plantations occupied 543,399 ha in Sarawak (Department of Agriculture Sarawak 2005), and the total area grew to

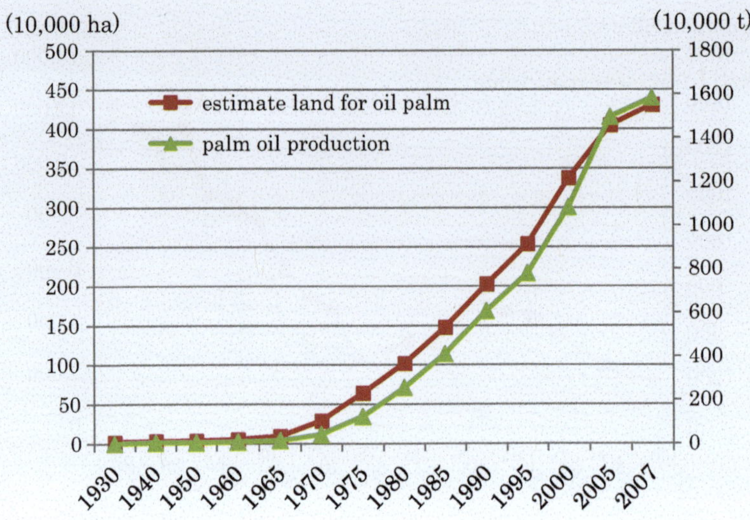

Fig. 10.2 Estimated land and production of palm oil in Malaysia. *Source*: Department of Statistics Malaysia, Malaysia Palm Oil Board

919,418 ha in 2010, representing a 1.7-fold increase over 5 years. In 2009, private companies farmed 82 % of the cultivated land of Sarawak, the government 13 %, and smallholders 5 % (MPOB 2010).

Recently, cultivated areas have expanded in other coastal areas, such as the Sibu and Bintulu Divisions. Plantation development in inland areas was late compared to coastal areas because of the inconvenient access to oil mills and inadequate road infrastructure. However, it is currently spreading to the inland hill areas, and oil palm cultivation by smallholders is diffusing to their post-swidden lands.

10.3 Migration from Riverside Longhouses to Roadside Settlements

From the 1970s, logging roads expanded from coastal to inland areas. Associated with this expansion, public road networks pushed into inland areas. Nowadays, roads cover all of Sarawak, and rural areas are directly connected to urban areas, a development which affects the spatial mobility patterns of the population. Previously, people lived in longhouses and cultivated their farms along rivers. Because of the hilly and mountainous topography of inland areas, rivers were the only means of transportation. However, after road networks were developed, people started to farm the land along roads. Recently, most of the people living in longhouses along rivers in the coastal areas have moved and reconstructed their longhouses along roads.

The same phenomenon exists in upper Rajang River of inland Sarawak, Malaysia. It is easier for people to use motorcycles than boats to access their farms.

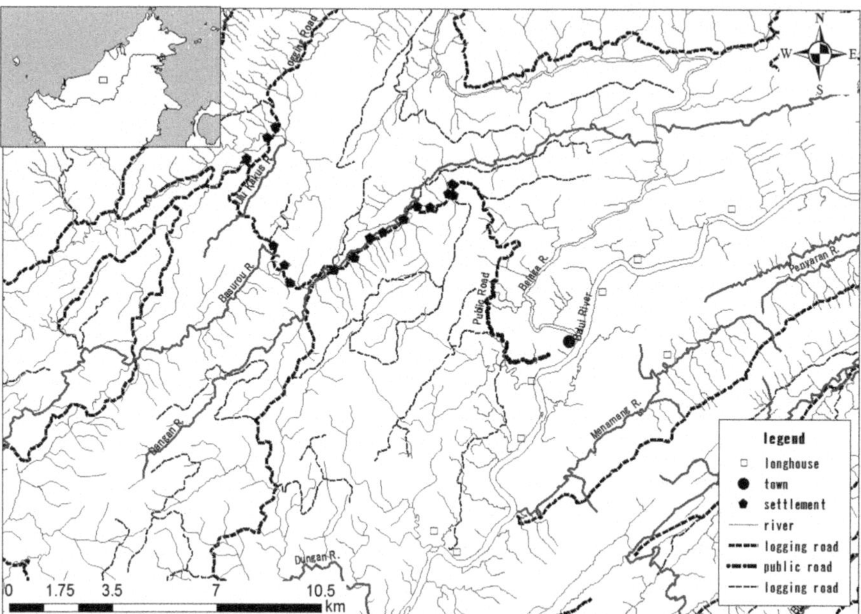

Fig. 10.3 Study Area

People have lived in scattered longhouse villages along the river. Most of these villagers have been swidden farmers. However, they started to cultivate paddy along the roads after road networks spread. The young generations have also migrated from inland riverside longhouses to urban areas because of the increasing availability of work in urban industrial areas (Soda 2007; Ichikawa 2011). The population outflow from longhouses to cities can be seen since the 1980s. Recently, the spatial patterns of inland population mobility expanded including riverside longhouses, urban areas, and roadside settlements. The frequency of their mobility also increased due to the use of motorcycles and cars between these plural residential sites.

I conducted my research in X village in upper Rajang River in 2008 and 2009. The people of X village originally lived in conventional longhouse located along a river. They gradually began to move to town because of the demands of construction works since the 1980s. Some of them also began to move to roadside settlements in the late 1990s, following other longhouse villagers who had moved earlier (Fig. 10.3). They established settlements and cultivated paddy along the road, a 1-h motorcycle trip from the town. More villagers joined them in the following years. They moved, first, because of insufficient primary forest around the conventional longhouse and, second, because of good accessibility from the town. They can gain access to their roadside farming sites by using motorcycles and cars from the town. It was more convenient for those who worked in the town, because they could engage in both agricultural work and waged labor. Recently they cultivated farms either near the longhouse or roadside settlements. The population flow from riverside longhouses to settlements along the roads greatly affected resource use and subsistence activities.

10.4 Changes in Local People's Resource Use and Subsistence Activities

How then have the spread of road networks and expansion of plantations affected people's subsistence activities? What are the differences in the subsistence activities of conventional riverside longhouse and new settlements along roadsides?

The historical change of people's subsistence activities in the case village is described below (Fig. 10.4). For a long time their lives basically relied on hunting and gathering in nearby forests, although they have also engaged in shifting cultivation and cash crop cultivation since the 1960s. Beside these basic subsistence activities, they also engaged in trade of several forest products. The international trade of forest products such as dammar and jeluton (*Dyera costulata*) resin was conducted until 1970s. However, economic impact of these trades was relatively low at that time. In contrast, the international trade of illipe nut and rattan cane flourished in 1970s and 1980s. It was continued until the 1990s. The cash income by these forest products was very important at that time. In addition, they also trade several forest products in a local market. It continues until today. They trade river fish, rattan handfcart such as basket or mat, wild plants such as bitter bean (*Parkia speciosa*) and rattan shoots, and fruits such as dorians in the local market. These trades are still main sources of their cash income. How do recent land use changes affect the local people's resource use and subsistence activities?

The vegetation around the longhouse consists of secondary forest and shifting cultivation. In comparison, the settlements along roads have been surrounded by oil palm and acacia tree plantations since 2005. Below, I describe the differences of resource use and subsistence activities between riverside longhouse and roadside settlements.

10.4.1 *Shifting Cultivation*

Shifting cultivation has been one of the main subsistence activities for them. Before the 1990s, most of the local residents opened primary and secondary forests around

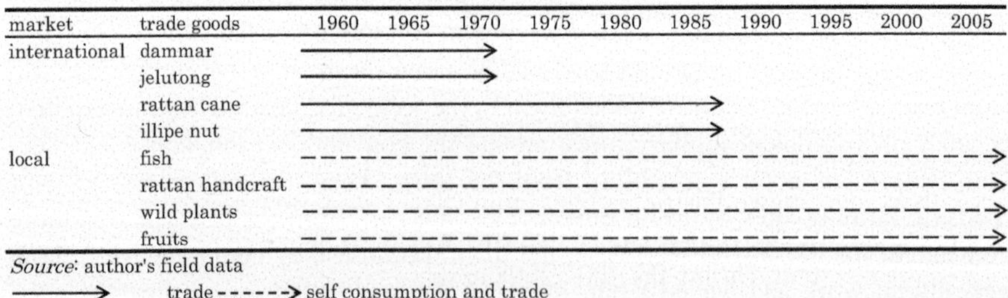

Fig. 10.4 History of forest product trade in the case village

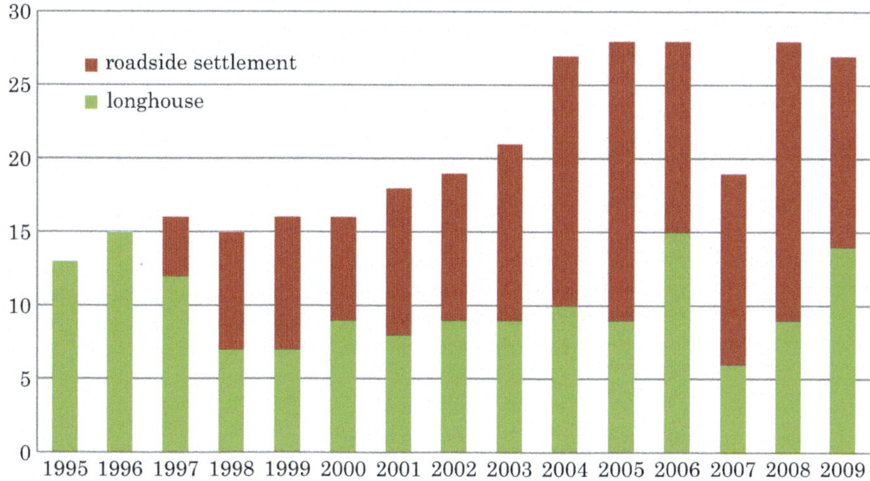

Fig. 10.5 Household number according to shifting cultivation sites. *Source*: author's field data

their riverside longhouse, where they cultivated paddy, corn, cassava, and other vegetables. However, after the expansion of road networks, people in the late 1990s moved to the roadsides, where their farming sites also shifted. In the X village I have recorded the farm sites of the villagers. Each household cultivated paddy, either near the longhouse or in the forest along the roads. In 1997, 12 households cultivated farms around the longhouse and only 4 households worked land along the road. In contrast, in 2008, only 6 households cultivated land around the longhouse and 13 households farmed along the roads (Fig. 10.5). Recently, households prefer to cultivate land along the roads.

The primary forests around the riverside longhouse had been cultivated for long time, so they claimed the soil became very poor. As seen in the figure, the numbers of households who cultivated their land along the logging roads increase from the 1990s. However, I found the crops in the two farming sites are almost the same. The average crop yield per household around the longhouse during the 11 years from 1997 and 2008 was 959 kg, and 955 kg along the roads. There were no clear differences in the process of cultivation. The people organized labor by exchange parties in both places. The area of cultivation land in both areas was almost the same. The vegetable species they cultivated differed only slightly (Table 10.1). They cultivated a larger variety of vegetables around the longhouse.

Villagers seem not to choose their farming sites based on crop differences but rather on the convenience of access to such sites. When they cultivate around the longhouse, they need to thresh crops manually. Furthermore, they must carry harvested crops on foot to the longhouse. It is also inconvenient for the households that engage in wage labor in town to carry the rice frequently from their longhouse. It is difficult to balance farming around the longhouse with work in the town. In contrast, people can use motorcycles to access their farming sites along the roads;

Table 10.1 Comparison of cultivated plants of longhouse and roadside settlements

Common name	Local name	Scientific name	Longhouse	Roadside settlement
Cassava	Kele	*Manihot esculenta*	+	+
Banana	Balat	*Musa* spp.	+	+
Amamesiva	Situn	*Sauropus androgynus*	+	+
Chili	Be	*Capsicum frutescens*	+	+
Sweet potato	Abai	*Ipomoea batatas*	+	
Taro	Sukai	*Colocasia esculenta*	+	
Lemon glass	Serai	*Cymbopogon citratus*	+	+
Galangar	Kua	*Languas galanga*	+	
Long bean	Beleta'	*Vigna sesquipedalis*	+	+
Dishcloth gourd	Kusut	*Luffa cylindrica*	+	+
Papaya	Medung	*Carica papaya*	+	+
Sugger cane	Tovu	*Saccharum officinarum*	+	+
Jack fruits	Badu'	*Artocarpus heterophyllus*	+	
Cacao	Koko	*Theobroma cacao*	+	
Lime	Limau	*Citrus microcarpa*	+	+
Pineapple	Aro'san	*Ananas comosus*	+	+

Source: author's field data

therefore the work is more effective. Furthermore, they can easily carry harvested crops by motorcycle and car to the settlement and to family members who live in town. These factors increase the household who cultivate their farms along the roadside recently.

10.4.2 Oil Palm Smallholding

The largest difference in the subsistence activities of the two sites concerns oil palm smallholdings, which are only available along the roads. When an oil palm plantation spreads around the village, some villagers started planting oil palm as smallholders. Oil palm smallholdings are becoming a significant subsistence activity along the road.

Although the people of Village X have engaged in the cultivation of various cash crops, none of these has become a stable subsistence activity. In the 1970s they were engaged in Para rubber trees (*Hevea brasiliensis*) cultivation. In the 1980s they cultivated coffee; in the 1990s, they undertook cacao and pepper cultivation (Table 10.2). However, none of these plantings provided sustainable cash income. Because of high prices and ease of cultivation, Cacao was once regarded as a golden crop. However, it was difficult to harvest because of the losses inflicted by squirrels in non-mass fruiting years in the case village.

With the road infrastructure, oil palm smallholding began. The smallholders must transport oil palm bunches to mills within 24 h of harvesting. People started

Table 10.2 Comparison of the number of households engaged in cash crop cultivation. The numbers on left sides indicate the number of plants of the households

Number of plants	Rubber	Coffee	Cacao	Pepper	Oil palm	Total
1–200	10	9	7	0	4	30
201–400	0	0	0	2	4	6
401–600	0	0	0	0	1	1
601–800	0	0	0	0	2	2
801–1,000	4	1	4	0	2	11
1,001–1,200	0	0	0	0	0	0
1,201–1,400	0	0	0	0	0	0
1,401–1,600	0	0	0	0	1	1
1,601–1,800	0	0	0	0	0	0
1,801–2,000	0	0	2	0	0	2
2,001–	0	0	0	0	1	1
Total	14	10	13	2	15	54

Source: author's field data

planting oil palms in 2008, following the lead of a farmer who lived near the settlement. When I conducted my survey in 2008, 58 % of the villagers were already cultivating them. They said they started planting oil palms because neighboring people had planted them and that they expected high incomes. Those who had not turned to the crop said that seeds and agricultural chemicals were too expensive, they were busy with other subsistence activities, or they were uneasy about growing and selling oil palms.

Most of the smallholders started planting palms within 10 years after the large scale plantation began operation in Sarawak. To start cultivation, they bought oil palm seeds from the nearby plantation company for 2.20 Malaysian Ringgit[1] (MYR) per seed in 2008. Of the 15 households who lived along the road, only 7 planted 50–750 of these oil palm seeds each, or 2,950 in total. The area they planted was usually less than 10 ha. Many people also took young sprouts from mother plants and grew them in a nearby gardens or planted seeds from fallen fruit. Palms grown from the latter are regarded as less fertile, but people with less funds still rely on them. The total number of palms in 15 households was about 5,000. Five household also planted jatropha (*Jatropha curcas*), 150 trees each, for a total a 750 trees.

Families did not open new secondary forest areas for oil palm gardens. They planted oil palms in paddy fields after harvests. Palm fruits are harvested within 3–4 years after planting. Therefore, in the next few years, oil palm smallholdings may become a more significant subsistence activity in this area if the current high market price is maintained. It may also attract people who live in riverside longhouses to turn to becoming oil palm smallholders in roadside locations. Within their

[1] One Malaysian Ringgit is about 0.3 US dollars.

subsistence activities, oil palm smallholding is a newly emerged option after the road connection was developed. It is also related to the current expansion of large scale plantations through out Malaysia.

10.4.3 Hunting

Hunting is one of the most important subsistence activities of the people in X village. Historically, people use various hunting methods, which are adapted to the habits and the characteristic of animals. Before the 1960s they used spears, blowpipes, spring traps, bamboo spears, and important game, such as bearded pig, barking deer, and sambar deer, were usually hunted using spears. They often took hunting dogs along with them to track and chase animals in the forest. The blowpipe was next in importance to the spear. They shot birds, and arboreal animals such as monkeys and tupai with blowpipes. Along with the spear and blowpipe, the local people employed spring traps to catch smaller mammals and birds, which were also snared in birdlime. Hunters had traditionally employed bamboo spears, which they don't use nowadays. Today, hunting has greatly changed. Hunters now use vehicles and guns to hunt animals, and hunting ranges are wider; they also hunt in the forest along roads.

Hunting methods, times, and target animals are no longer the same. When still living in longhouse, people generally hunted during the day, but now they hunt at night with torchlights in forests near plantations because wild boar frequently come to feed on oil palm fruit in the plantations along the roads only at night. In addition, they find hunting around the plantations more convenient, since they can use motorcycles on the roads. Moreover, many people stated that hunting around the plantations is easier than hunting in the natural forests. In fact, field data from 2008 reveal that more than half of all successful hunting took place around plantations. People hunted the bearded pig, mouse deer, or barking deer near longhouse, but now search mainly for bearded pig around the plantation.

10.4.4 Gathering Wild Plants

In the focal village, the gathering of wild plants has been very important for villager livelihoods for a long time. In total, the villagers have used 67 species of wild plants and fungi around the longhouse, including three species of the sago palm (*Arenga undulatifolia*, *Eugeissona utilis*, *Metroxylon sagu forma longispinum*), which they gathered in the secondary forest as staple foods. Forty species were employed as non-staple foods, 23 species for snacks, and a single species for drinks. The non-staple food included 15 species of fungi, 6 species of ferns, and 4 species of ginger, all gathered in the secondary forest near the longhouse.

Table 10.3 Comparison of edible wild plants of longhouse and roadside settlements

Common name	Local name	Scientific name	Longhouse	Roadside settlements
Gnemon tree	Sabung	*Gnetum genemon*	+	
Bamboo shoot	Tuvu'	*Dendrocalamus* sp.	+	
Red gingerwort	Tipu'	*Etlingera punicea*	+	+
Marang	Tahap	*Artocarpus odoratissimus*	+	
Aren gelora	Jemako	*Arenga undulatifolia*	+	
Durian burong	Lubo'	*Durio* sp.	+	
Kadjatoa	Bulong	*Eugeissaona utilis*	+	
Bitter bean	Pata	*Parkia speciosa*	+	
Rattan shoot	Poha'	*Calamus* sp.	+	
Blimbi	Belibit	*Averrhoa bilmbi*	+	
Johore jack	Lumu'	*Artocarpus odoratissima*	+	
Turkey berry	Ulom	*Solanum torvum*	+	
Torch ginger	Uhom	*Etlingera elatior*	+	+

Source: author's field data

Along the roads around the plantations the villagers continue shifting cultivation and hunting, but they do not collect wild plants, except for rattan, since many plant species are difficult to find there. Table 10.3 shows the 14 species of edible wild plants which were most used in longhouse. Only two species are used in the settlement along the road. It is inconvenient for them to collect edible wild plants along the road. There are several reasons that they cannot use edible wild plants along the road. First, the villagers claim the vegetation around the riverside longhouse and road side settlements are different. For instance, there are plenty of sago palms around the longhouse, however we cannot find any sago palms along the road. Second reason is the access to the forest. The longhouse is surrounded by forest. Therefore they can easily take wild edible plants in their daily diet. In contrast, the roadside settlements are surrounded by oil palm plantation. Therefore the distance to forest is farther. Access to the forest is also lesser compare to the life in longhouse. The third reason is effect of herbicides. Villagers say they are afraid of taking edible wild plants around the plantation. They think these plants are harmed by herbicides sprinkled in the plantation and not safe to eat. These reasons are related to the differences in the use of edible wild plants.

The only useful plant around roadside settlements is rattan, plentiful in primary forests around plantations but exhausted near longhouse due to decades of gathering (Figs. 10.6 and 10.7). Currently, people who live around plantations engage more in collecting rattan and making rattan handcrafts than those who live in the longhouse. The cash income generated by selling rattan handcrafts is very important for women. In both residential sites, 22 women make and sell rattan handcrafts; 16 of these women live along the road. The highest cash income from selling rattan handcrafts was MYR 780 a month, although, on average, the women earned MYR 278 per month. Figure 10.8 provides a comparison of the rattan handcrafts of women living in the longhouse (6 women) and women living in roadside settlements (16 women). Six handcraft weavers live in the longhouse, and each

Fig. 10.6 A woman collecting rattans in the forest around the longhouse

Fig. 10.7 People collecting rattans in the forest along the road. (They can collect more rattans in the forest along the road)

earns an average of MYR 257 per month from the sale of handcrafts. In contrast, 16 weavers live in roadside settlements and earn MYR 369 on average per month per person. Compare to wild plants gathering around the longhouse, they can hardly collect edible wild plants, however, they can collect rattans more actively around the road side settlements.

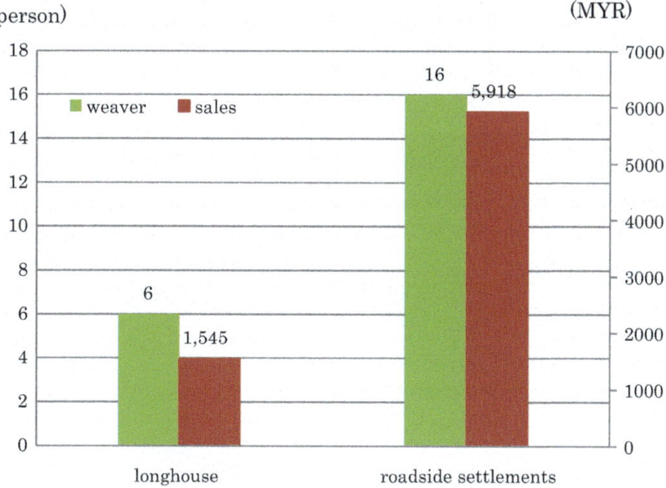

Fig. 10.8 Number of weaver and Sales of rattan handcrafts in longhouse and roadside settlements. (Total weaver were 22 people in both sites. Sales data was collected in June, November, and December 2008). *Source*: author's field data

Subsistence activities of X villagers in the longhouse and around the plantation in roadside settlements differ. While paddy cultivation and hunting are expected to continue in both sites, villagers cannot fish and collect edible wild plants around the plantation in roadside settlements. They can fish easily around the longhouse, because longhouse is located at riverside. In contrast, they can hardly fish around the settlement along road, because river is far from the settlement. In addition, they are afraid of the effect of herbicide flowing from plantation. However, they engage in hunting and gathering wild rattan more actively in the forests along the road. Furthermore, they plant oil palm on smallholdings.

10.5 Discussion

Palm oil production is becoming one of the most important industries in Insular Southeast Asia (Cramb and Curry 2012). Therefore, oil palm plantations are expected to expand continuously. In this situation, oil palm plantations become premise of local people's life. Local people also started to earn certain profit by planting as smallholder recently (Feintrenie et al. 2010; Kato and Soda 2012; Cramb and Sujang 2013). What is important here is how local people can control the stable livelihood under this situation. Another important background concerning to this point is expansion of road networks. The expansion of road networks changes local people's resource use. It became road based resource use.

This study considered the differences in the resource use and subsistence activities of the people who lived in conventional longhouse, surrounded by secondary forests along rivers, and those who dwelled in new settlements, surrounded by oil palm plantations along roads.

Subsistence activities and resource use differ according to the vegetation around settlements. There are vast secondary forests around the longhouse as a result of hill paddy cultivation. Even where some primary forests remain, they are far from longhouse or unfavorably placed. Under these conditions, people engage in shifting cultivation in secondary forests, where they gather wild plants, including many edible varieties, hunt, and fish. However, they do not collect rattan actively because these sites are exhausted. Similarly, they cannot plant oil palms around longhouse, given the lack of roads.

In contrast, the vegetation along roads consists of primary forest, secondary forest, and plantations. Although people cannot fish or collect edible wild plants, they can engage in shifting cultivation, hunt, and collect rattan. The people of village X regard hill paddy cultivation in the primary forest around the plantations as convenient because they can use motorcycles to transport harvested crops. They also claim that gathering rattan in the primary forest here is easier than in the forest around longhouse, where the soil is exhausted. The villagers think that hunting is easier near plantations because they can use motorcycles for transportation. In addition, they plant oil palms as smallholders. Thus, there are more livelihood alternatives along the road. However, primary and secondary forests remain very important, even for people living along road. If there is no forest around the plantation, they cannot continue paddy cultivation, hunting, and collecting rattan. The alternatives of several subsistence activities are very important for their stable livelihood. In coastal areas in Sarawak, some villages are more strongly affected by oil palm industry (Cramb 2011). In these areas, most of the villagers do not continue paddy farming, because oil palm smallholding becomes primarily important economic activities (Kato and Soda 2012). However, in inland Sarawak, the access to the palm oil mill is rather difficult, therefore they cannot concentrate only in oil palm smallholding. In this situation, the alternatives of other subsistence activities become comparatively significant.

The expansion of road networks in inland areas will continue to push the population from riverside longhouses to roadside settlements and urban areas. Conventionally, social space in Sarawak was centered around rivers, which is why it is called a riverine society. People built their longhouses and opened paddy fields along rivers. However, the expansion of logging into inland areas expanded road networks and the expansion of the oil palm industry has pulled the inland population to the roadsides. The expansion of road networks also changes local people's access to natural resources. They become able to access to new resources like rattans when the road networks arrive in the areas where they could not access by river based transportation before.

Urbanization is one of the other factors in this dynamic. Once inland areas were connected to coastal cities directly by road, access to the latter became easier. Young people were able to seek work in urban areas, and people moved more frequently between longhouses and cities. These population flows are also related to resource use changes along roadsides.

Oil palm smallholdings have become very common in inland areas. If palm oil prices remain high, smallholdings will continue to prosper; if they fail, the population will have to turn to other economic activities. In this sense, multiple subsistence activities, based on the forest, bolster local livelihoods. However, resource use along roadsides will become more general as the road network increases in density. In this situation it is necessary to continue to consider the stable subsistence model, including the livelihood along roadside settlements.

Acknowledgments I thank Prof. Shoko Sakai and the Sarawak Development Institute for their considerable support for this study. I also thank the GIS unit of the Forest Department of Sarawak, Dr. Reiichiro Ishii, Dr. Izuru Saizen, and Dr. Hiromitsu Samejima for providing a base map of the study area.

References

Barney K (2004) Re-encountering resistance: plantation activism and smallholder production in Thailand and Sarawak, Malaysia. Asia Pac View 45(3):325–339

Cramb RA (2011) Reinventing dualism: policy narratives and modes of oil palm expansion in Sarawak, Malaysia. J Dev Stud 47(2):274–293

Cramb RA, Sujang PS (2011) "Shifting ground": renegotiating land rights and rural livelihoods in Sarawak, Malaysia. Asia Pac View 52(2):136–147

Cramb R, Curry GN (2012) Oil palm and rural livelihoods in the Asia-Pacific region: an overview. Asia Pac View 53(3):223–239

Cramb RA, Sujang PS (2013) The mouse deer and the crocodile: oil palm smallholders and livelihood strategies in Sarawak, Malaysia. J Pheasant Stud 40(1):129–154

De Koninck R (2011) Southeast Asian agricultural expansion in global perspective. In: De Koninck R, Bernard S, Bissonnette JF (eds) Borneo transformed: agricultural expansion on the Southeast Asian frontier. NUS Press, Singapore, pp 1–9

Department of Agriculture Sarawak (2005) Agricultural statistics of Sarawak. Department of Agriculture Sarawak, Kuching

Department of Statistics Sarawak Year book of Statistics 1970, 1975, 1985

Department of Statistics Malaysia Year book of Statistics 1970, 1980, 1990, 2000

Feintrenie L, Chong WK, Levang P (2010) Why do farmers prefer oil palm? Lessons learnt from Bungo district, Indonesia. Small Scale Forest 9(3):379–396

Forest Department Sarawak Annual Report 1965–1985

Ichikawa M (2011) Factors contributing to different rates of depopulation in rural villages in Sarawak. Borneo Res Bull 42:275–288

Kato Y, Soda R (2012) Emerging trends in oil palm smallholdings in Sarawak, Malaysia. Geogr Stud 87(2):26–35 (in Japanese)

Majid Cooke F (2002) Vulnerability, control and oil palm in Sarawak: globalization and a new era? Dev Change 33(2):189–211

Malaysia Palm Oil Board (2010) Malaysian palm oil statistics. Malaysia Palm Oil Board, Kajang

Ngidang D (2002) Contradictions in land development schemes: the case of joint venture in Sarawak, Malaysia. Asia Pac View 43(2):157–180

Ooi KG (1995) An economic history of Sarawak during the period of Brooke rule, 1841–1946. Dissertation, Hull University

Smythies BE (1963) History of forestry in Sarawak. Malayan Forest 26:232–253
Soda R (2007) People on the move: rural–urban interactions in Sarawak. Kyoto University Press and Trans Pacific Press, Kyoto and Melbourne
Tagliacozzo E (2005) Onto the coasts and into the forests; ramifications of the China trade on the ecological history of Northwest Borneo, 900–1900 CE. In: Wadley RL (ed) Histories of the Borneo environment. KITLV, Leiden, pp 25–59

Index

S. Sakai and C. Umetsu (eds.), *Social–Ecological Systems in Transition*, Global Environmental Studies, DOI 10.1007/978-4-431-54910-9, © Springer Japan 2014

The manufacturer's authorised representative in the EU is Springer Nature Customer Service Centre GmbH, Europaplatz 3, 69115 Heidelberg, Germany. If you have any concerns regarding our products, please contact ProductSafety@springernature.com

Printed and bound by CPI Group (UK) Ltd, Croydon, CR0 4YY
15/07/2026
02167657-0001